# FISH FERMENTATION

TRADITIONAL TO MODERN APPROACHES



NIPA® GENX ELECTRONIC RESOURCES & SOLUTIONS P. LTD.

New Delhi-110 034

www.nipaers.com
Browse, Search, Read & Buy...
Print Books
eChapters
eBooks
Publishers
Forthcoming
Subject Catalogues
New Titles
Online Resources on:
✓ Current Affairs
✓ Reasoning, Logic & Aptitude
✓ Competitive Examinations
✓ English Language Lab
✓ Writing & Pronunciation Tools
✓ Personality Development
✓ Online Programmes for Professionals
✓ Interview Preparation
Pay Using
PhonePe
Google Pay
paytm
PayPal
UPI
CC Avenue

# Fish Fermentation

## TRADITIONAL TO MODERN APPROACHES

*by:*

**DEBABRAT BAISHYA**
Lecturer
Department of Biotechnology
Pandu College, Assam

**MANAB DEKA**
Reader
Department of Biotechnology
Gauhati University, Assam

**NIPA® GENX ELECTRONIC RESOURCES & SOLUTIONS P. LTD.**
New Delhi-110 034

**NIPA® GENX ELECTRONIC RESOURCES & SOLUTIONS P. LTD.**

101,103, Vikas Surya Plaza, CU Block
L.S.C. Market, Pitam Pura, New Delhi-110 034
Ph : +91 11 27341616, 27341717, 27341718
E-mail: newindiapublishingagency@gmail.com
Web: www.nipabooks.com

***For customer assistance, please contact***
Phone: + 91-11-27 34 17 17
Fax: + 91-11- 27 34 16 16
E-Mail: feedbacks@nipabooks.com



ISBN: 978-81-19103-83-6



Composed and Designed by NIPA®.

# PREFACE

Incredible advances are occurring in the field of Industrial Microbiology at a dizzying pace, and modern technology such as biotechnological tools has made an impact on various aspects in day today life. ***Fish Fermentation: Traditional to Modern Approach*** is the first of its kind as this book has introduced both the traditional and modern methodologies of an important food commodity i.e. fish. The contents of the book are aimed at a diverse group of students ranging from undergraduate to postgraduate. Furthermore, it will help research students in the discipline of fermentation, in particular, solid substrate fermentation in understanding the traditional fermentation along with the microbial processes involved in such fermented products. It goes without saying that there are a good many number of books on Food Microbiology but the unique feature in this book is that students will get a chance to know the gathered information on fish fermentation in a single volume that we feel a need of the hour for the students. Chapter 1 to Chapter 4 will act as basic guideline for the student by proving more important facts about the traditional process of fermentation of fish, different fermented fish products in different parts of the World, their microbial diversity and nutritional chemistry. Chapter 5 and Chapter 6 includes the modern, cutting-edge, emerging technologies that has been introduced in fermenting fish with experimental detail and contain basic and advances in the field of probiotics research. The strategy in selecting starter culture for fish fermentation will no doubt help the research students in designing their work with modified novel technique in this discipline.

Further, have been gratified to hear from instructors and students for making it more suitable and for any unknown mistakes.

**Debabrat Baishya**
**Manab Deka**

# CONTENTS

Chapter 1

# Fermented Food Products

Fermented foods are food substrates that are produced due to the cumulative activity, usually of edible microorganisms, whose enzymes, particularly amylases, proteases and lipases hydrolyze the polysaccharides, proteins and lipids to non-toxic products with flavours, aromas and textures pleasant and attractive to the human consumer. The role played by fermentation during food processing helps in - enrichment of human dietary through development of wide diversity of flavours, aromas and textures in food; preservation of substantial amounts of food through lactic acid, alcoholic, acetic acid, alkaline fermentation and high salt fermentations; enrichment of food substrates biologically with vitamins, protein, essential amino acids, essential fatty acids and antibiotics; detoxification during food fermentation processing and reducing the cooking times and fuel requirements.

Food fermentation can be classified in a number of ways (Dirar, 1993): **by categories** (Yokotsuka, 1982) - (a) alcoholic beverages fermented by yeasts, (b) vinegars fermented with *Acetobacter*; (c) milks fermented with lactobacilli, (d) pickles fermented with lactobacilli, (e) fish or meat fermented with lactobacilli, and (f) plant proteins fermented with molds with or without lactobacilli and yeasts; **by classes** (Campbell-Platt, 1987) - (a) beverages, (b) cereal products, (c) dairy products, (d) fish products, (e) fruits and vegetable products, (f) legumes, and (g) meat products; **by commodity** (Odunfa, 1988) - (a) fermented starch roots, (b) fermented cereals, (c) alcoholic beverages, (d) fermented vegetable proteins, and (e) fermented animal protein. Steinkraus (1983, 1996) classified fermentation as - (a) fermentations producing textured vegetable protein meat substitutes in legume/cereal mixtures, (b) high salt/savory meat flavoured/amino acid/peptide sauce and pate fermentations, (c) lactic acid fermentations, (d) alcoholic fermentations, (e) acetic acid/vinegar fermentations, (f)alkaline fermentations, (g) leavened breads, and (h) flat unleavened breads.

Religion was an attempt by humans to explain the unexplainable origin of the universe, the earth and man long before there was a scientific method or the means to study these difficult problems and no concept of, for example, microorganisms, knowledge of which we obtained only about 300 years ago when Leeuwenhoek discovered tiny animalcules under his primitive lenses and only a little more than hundreds years ago when Pasteur demonstrated the role of microorganisms in fermentation and Koch showed that microbes causes disease. And it is only in the last over 50 years that knowledge of the role polymeric deoxyribonucleic acid plays in all forms of life was discovered. According to the present scientific thought, the earth is about 4.5 billion years old. The first forms of life to appear or evolve on earth were microorganisms. Since then and till today, microorganisms have had and have the principal task of recycling organic matter in the environment. Whether it was by chance or by design, it was extraordinarily fortune that the earth was originally colonized by microorganisms. Early men very likely consumed fruits, leaves, berries, seeds, nuts probably tubers foraging from place to place as apes do today. There was relatively large potential food supply and relatively few humans. Excess food supplies, fruits, berries, etc., fell on the ground and the seeds either germinated or the carbohydrates, proteins, fats, and so on, were consumed by microorganism using enzymes that converted fermentable carbohydrates to alcohol or acids and finally to water and carbon dioxide. As population increased it became desirable for humans to collect fruits, etc. as a store of food to tide humans over during periods of bad weathers when fresh food was not readily available. During storage foods were invaded by microorganisms producing toxins making the food unsafe for consumption. When the products of invasion are ill-smelling, off-flavoured or toxic, human consumers try to avoid them and foods are described as spoiled. If the microbial products are pleasantly flavoured, have attractive aromas and textures and are nontoxic, the human consumer accepts them and they are designated as fermented foods (Steinkraus, 1996).

Fermentation technology is one of the oldest food technology applications that has been developed and utilized for survival. The origin of Asiatic fermentation technology evolved as early as the littoral forager's period of the Primitive Pottery Age (8,000-3,000 B.C.), which led the Neolithic culture of agriculture in the Northeast Asia. Although, fermentation (e.g., brewing and wine production) was done for many hundred years, yet during the end of 15th century, brewing became partially industrialized in Britain. By early 19th century Cagniard-Latour (1837) and Schwann (1837) reported that the fermentation of wine and beer is accomplished by yeast cells. It was Luis Pasteur who observed association of microorganisms in the process of fermentation. It seems that the art of fermentation originated in the Indian subcontinent, in the settlements that predates the great Indus Valley civilization. During the Harappan spread or pre-Vedic times, there are indications of a highly developed system of agriculture and animal husbandry. Artifacts from Egypt and the Middle East also suggest that fermentation was known from ancient times in that region of the world (Upadhyay, 1967). As a process, fermentation consists of transformation of simple raw materials into a range of value-added products by utilizing the phenomenon of growth of micro-organisms and/or their activities on various substrates (Joshi and Pandey, 1999). This means that knowledge of microorganisms is essential to understand the process. In fact, when we study fermented foods, we are studying the most intimate relationships among humans, microbes and foods.

In traditional method of fermentation it is an observable fact that the raw materials used in the fermentation contain both desirable and associated microorganisms. Again technique like back slopping is more common in the traditional fermentation process where some products from a successful fermentation are added to the starting material and conditions are set to facilitate the growth of microorganisms that has been introduced from the previous product. In case of fish fermentation also as has been practiced by different communities of the people all over the world, the scenario is no different. Fermentation of fish by natural fermentation or back slopping is not hygienic and totally uncontrolled system, which allows the contamination of such product with pathogenic organisms either from the untreated partially cured fish or from improper handling during processing. Such conditions increases the food borne illness producing so called the biological hazards and consumption of such food product is quite unsafe. However, Steinkraus (2002) reported that fermented foods generally have a very good safety record even in the developing world where the foods are manufactured by the people without training in microbiology or chemistry, in unhygienic and contaminated environments. The fermentation of a variety of foods, involved mainly the production of lactic acid where the lactic acid organisms converts the fermentable sugars into lactic acid. This single category is responsible for processing and preserving vast quantities of human food and ensuring safety.

The nutritional impact of fermented foods on nutritional diseases can be direct or indirect. Food fermentations that raise the protein content or improve the balance of essential amino acids or their availability will have a direct curative effect. Similarly fermentation that increase the content or availability of vitamins such as thiamine, riboflavin, niacin or folic acid can profound direct effects on the health of the consumers of such foods (Jelliffe, 1968).

The diverse ethnic group in the North East region of India making it opulent with their variegated ethnic composition is popular for its own brand of indigenously processed food and beverages. A total of seven states viz. Assam, Arunachal Pradesh, Nagaland, Manipur, Mizorum, Meghalaya and Tripura comprises the north eastern part of India with diversified geomorphological and agroclimatic conditions from tropical to temperate to alpine. Recently another state Sikkim has also been introduced as a member of this region. A large number of tribes inhabit the hills and plains, entered this region at different points of time and through different routes after which they occupied marginal territories and living for over centuries in isolation. They developed their own socio cultural pattern which also included a distinct food habit. Besides rice and pulses which are the staple foods of this region, the people of this region from time immemorial had been consuming fermented food products prepared from a variety of sources such as rice, fruits, vegetables, fish and meat. Almost all major ethnic groups of this region have their own indigenous technology for improving the qualities and self-life of certain food through natural fermentation. These technologies are therefore, classical examples of traditional biotechnology processes where some domesticated microbes are utilized for producing certain improved food items.

The fermented food and beverages of the northeast India are not commercialized like other Indian items namely soy-sauce, cheese, idly, yoghurt and many alcoholic beverages. The primary reason being the lack of standard process of production. There are many such fermented food and beverages which are not even known to the people from various states of the northeast region. **Soybean** is widely cultivated and the seeds are regularly fermented by traditional method in Arunachal Pradesh, Manipur, Nagaland, Mizorum and Meghalaya. Indigenous fishes seasonally available in large quantities are fermented in Meghalaya, Mizorum, and Manipur. Fermented **Colocasia** is a product from Nagaland which has got a long shelf-life but not popular in other states. Fermented mustard seeds commonly known as **Kharoli** are not common among the people from hilly states but are a common pickle item in Assam. Similarly submerged fermented rice (**Poita bhat**) is a common cereal based fermented product from Assam. Rice based alcoholic beverage is popular through out the region. For this, the fermentating microbes are domesticated and used in the form of "starter culture". Fermented young bamboo shoots (**Khorisa**), fermented leafy vegetables (**Gundruk**) etc. are some more examples from Assam.

In India, particularly in North Eastern region, fish is the most favourite food item. Fishes, as they are highly perishable, are cured as a means to preserve them. The methods of preservation also provide cultural identity amidst the ethnic groups. Fishes are consumed as fresh, partially cured or fermented forms. Partially cured and fermented fishes allow the constant availability of fish throughout the year. During fermentation it is observed that the raw material undergoes extensive transformation of organic substrate. The process of fermentation in fish is carried out initially by aerobic followed by microaerophillic and then by anaerobic bacteria; the enzyme present in the fish has major role in the process during which the fish autolyse largely through the action of enzymes.

Fish and fish products among the others play a major role in human as a source of biologically valuable proteins, fats and fat soluble vitamins. Fish is digested substantially faster and therefore, have a much lower nutritive saturation value. Preservation of nutritive value of fish is very important as the rigor mortis is much shorter in them and hence starts degrading very fast soon after its death. Further, fish being a low acidic food, is very susceptible to the growth of food poisoning bacteria. Drying and other curing techniques are most popular techniques of preservation in developing countries due to their economic viability.

It is evident that fermentation enhances the hedonic values of the food products by certain bio-chemical changes brought about by the endogenous enzymes secreted by the microorganisms. A number of factors contribute to the unique personality of fish flesh. Fish lipids contain a high proportion of polyunsaturated fatty acids which are more reactive chemically than the largely saturated fats that occur in mammalian meat. This makes fish far more susceptible to the development of oxidative rancidity. Fish flesh naturally contain very low levels of carbohydrate and these are further depleted during the process of fermentation as the microbes present in the fish use these carbohydrate for their metabolic activity. Fish protein occupies an important place in human nutrition. They have high digestibility and biological value and growth promoting values. They are well balanced with respect to essential amino acids and are comparable to other proteins of animal origin. The composition of non-protein nitrogen fraction of fish also plays a significant role in bringing the putrefactive changes in fish flesh during fermentation.

The state of Meghalaya is situated between $25^0$00′ and $26^0$10' north latitude and $89^0$45′and $92^0$47′ east longitudes and occupies an area of 22,429 sq. km. The state is bounded in the northeast and west by Assam and in south and parts of west by Bangladesh. It has got five administrative districts with Shillong as capital. The climate of the state is per-humid with small seasonal water deficiency. The average rainfall ranges between 2400 to 4000mm, confined mostly from April to October and scanty in the dry winter months. The mean summer temperature is $26^0$C and mean winter temperature falling down to $9^0$C. At higher elevations it drops below freezing point. The mean annual soil temperature at higher elevations is less than $22^0$C.

The 'Khasi' tribes of Meghalaya prepare a fermented food product out of a trash fish species *Puntius sarana* (Ham.) and the product is locally known as 'Tungtap'. Sun-dried *Puntius sarana* is observed to be comparatively stable with respect to other similar varieties of fish. Even when the fish is dried without gutting, it can be preserved for a long duration (almost up to one year) without much change. The factor contributing to this is low moisture content or water activity as a result of sun drying. Another factor may be the presence of antimicrobial agents that might be responsible for keeping quality of the product. The traditional belief of health related benefit of fermented fish may account for presence of such antimicrobial agents or the bioactive contents that may be produced by the microorganisms while dwelling within the fish during fermentation or curing process. The systematic position and habitat of this fish species is as follows –

| | | |
|---|---|---|
| Phylum | : | Chordate |
| Sub-phylum | : | Vertebrata |
| Super-class | : | Pisces |
| Class | : | Actinopterygii |
| Order | : | Cypriniformes (Carps) |
| Family | : | Cyprinidae (Minnows or carps) |
| Genus | : | Puntius |
| Type | : | *Puntius sarana* |
| Common name | : | Olive barb |
| Local name | : | Puthi (in Assam) |
| Environment | : | Benthopelagic, Potamodromous, freshwater, brackish |

Climate : Tropical

Distribution : Asia, Afghanistan, Pakistan, India, Nepal, Bangladesh, Bhutan, Sri Lanka and Thailand.

Sun-dried *Puntius sarana* is observed to be comparatively stable with respect to other similar varieties of fish. Even when the fish is dried without gutting, it can be preserved for a long duration (almost up to one year) without much change. The factor contributing to this is low moisture content or water activity as a result of sun drying. Another factor may be the presence of antimicrobial agents that might be responsible for keeping quality of the product. The traditional belief of health related benefit of fermented fish may account for presence of such antimicrobial agents or the bioactive contents that may be produced by the microorganisms while dueling within the fish during fermentation or curing process.

There is a transitional step between fresh fish and fermented fish which is termed as partially cured or semi cured fish. These partially cured fishes are came from different parts of the country to the wholesale market of Jagiroad, a place under Morigaon district of Assam from where they are brought to different places of the northeastern region for further processing by indigenous technique. Therefore, these partially cured fishes are used as raw material for fermentation process by the various ethnic groups of the region. Traditional fermentation is a low cost method of fish preservation using artisanal equipment which is readily available, easy to fabricate and repair. Therefore, one does not need a large capital outlay to start operations. In most cases, equipments such as old barrels, earthenware pots, old nets, locally-made drying racks, mats, used jute/poly sacks and cans which are the major items used, are locally available and affordable. Therefore it needs little capital to invest in order to commence business. While making a market survey on *Puntius sarana* in order to know the market potentiality it is found to be a better money rolling business. The survey revealed that the cost of partially cured fish range between Rs.100-150/- per Kg and it is Rs.300-350/- per Kg in case of fermented fish while it is Rs. 30-50/- per Kg in case of fresh fish. The increase in the prices of the processed fish is due to reduction in weight which is brought about by moisture loss during salting and/or drying. Curing results in about a 50-60 percent reduction in weight and the processors adjust the price accordingly. A total of 1440 tones of these products are sold per year in the whole-sale market of Jagiroad. Considering the lowest price @ Rs.100/- per Kg in case of partially cured fish the total turn out will be Rs.14,40,00000/- per year. Once fermented, the cost increases is nearly three times. Therefore, the projected cost @ Rs.300/- per Kg will be Rs. 43,20,00000/- per year. But in spite of having such a fascinating economic potential, the product is consumed by less than about fifty per cent people in the region. Moreover, the traditional process of fish fermentation is not upto the mark as far as the good manufacturing practices (GMPs) and hazard analysis and critical control point (HACCP) are concerned. It is seen that many processors use simple artisanal technologies for fermentation, packaging and storage. These methods usually date back in history or were introduced into country by settler communities. The methods are easily transferred either by tradition within a family or through non-formal training. As a result of the mode of technology transfer, there is a lack of standardization in production processes and product quality differs from batch to batch or from one locality to another.

This market survey and all other facts is therefore projecting the drawbacks that the lack of proper scientific documentation and lack of a standard processing technology are the main reason why such fermented food products of this region are not commercialized at par with to catch hold of the international market and henceforth scientific input to this indigenous food product is of utmost necessity so far a good health condition of the consumers are concerned and at the same time it will boost to uplift the market potential of this product at international platform.

Moreover, many indigenous traditional technologies are not easily adopted by transnational companies without altering the methods of preparation and perhaps ending up with a product of altered flavor and unacceptability. The international industries are likely to try to produce the most acceptable products accepted worldwide. To expand the market to western peoples who consume different food cultures, the producers might have to aim for indigenous fermented foods for health as a food recommended for its nutrition value and as such there is an enormous market worldwide waiting for Indian fermented food products. The opportunity of exporting these products is great with Indian dishes in the kitchen of the world. It is quite a challenge for both producers and Indian government to manage such industries towards internal markets. The use of development in biotechnology, standardized processes, including international laws and regulations to ensure quality standard and food safety will make possible for Indian fermented food products to smartly gain more market share in this competitive world of exports.

## References

Cagniard-Latour, C. (1837). *Ann Chim. Phys.* 68: 206-222.

Campbell-Platt, G. (1987). Fermented foods of the world: a dictionary and guide. London, UK, Butterworths, p290.

Dirar, H.A. (1994). The indigenous fermented foods of Sudan. Wallingford, Oxon, UK. CAB Int, p552.

Jelliffe, D. B. (1968). Infant Nutrition in the Tropics. World Health Organization (WHO). Geneva, Switzerland. p. 335.

Joshi, V. K. and Pandey, A. (1999). Biotechnology: Food Fermentation. Vol.1, Educational Publishers and Distributors. pp 1-24.

Odunfa, S.A. (1988). Review: African fermented foods: from art to science. *Mircen J.* 4: 225-273.

Schwann, T. (1837). *Ann. Phys. Chem.* (Poggendorff). 41: 184-193.

Staeikraus, T. (1983). Handbook of indigenous fermented foods. New York, NY:Marcel Dekker, p 671.

Steinkraus, T. (1996). Handbook of indigenous fermented foods, 2$^{nd}$ edition, revised and enlarged. New York, NY:Marcel Dekker, p776.

Steinkraus, K. H. (2002). Fermentations in World Food Processing. *Comprehensive Rev. Food Sc. Food Safety.* 1: 23-32.

Yokotsuka, T. (1982). Traditional fermented soybean foods. In: Fermented foods ed. Rose A.H., London,UK, Academic Press.

# Chapter 2

# Traditional Fish Preservation Techniques

It is an old practice that during the flourishing period, the fishes specially trash fishes are caught from their natural habitat and a bulk of them are preserved to make available for further use. Traditional fish preservation techniques are mainly based on combination of salting, drying and smoking. These techniques have allowed more widespread consumption of fish. Sun drying is most ancient method of fish preservation. Drying removes moisture from tissues and help to arrest bacterial and enzymatic putrefaction.

Fermented fish have, for many years, been considered as a Southeast Asian product. These products are highly salted and fermented until the flesh is transformed into simpler components. Fermentation of fish is a product in which the identity of the fish is maintained to large extent. Fermented fish products as fish sauces, fish paste, or salted fish have been consumed since ancient times. Because of poor roads and other methods of transport, the provision of fresh fish to potential inland consumers was impossible, and this encouraged fermentation as a preservation technique. In Southeast Asian countries such as Thailand, Kampuchea, Malaysia, the Philippines, and Indonesia, the use of fermentation as a preservation method for fish has been of great value since earliest times. In the countries of northern Europe, fermented fish products are used mainly as condiments, whereas in Southeast Asia, various fish products are regarded as staples. In India and Sri Lanka, pickled and in Colombo, cured fish have been known for many years (Beddows, 1985).

Different processes employed in the fermentation of fish yield three different types of product, which are –

- Products in which the fish substantially retain its original form.
- Products in which the original fish is reduced to the form of a paste.
- Fish sauce in which the meat is reduced to a liquid.

Fermented fishery products are generally classified into three categories:

- High salt (20% or more) : Liquid separated (fish sauce)
  : Residue (cured fish)
  : Mincing and partial drying (fish paste)
- Low salt (6-8%) : Lactic acid fermentation
  : Acid picking at low temperature

- No salt : Dried bonito fermentation
  : Alkaline fermentation.

(Gopakumar, 1997).

Most of the fermented fishery products are made from fatty fish. Lean fish has sometime been noted to give a less acceptable texture and flavour. Fish paste is such a product involving fermentation and processed in many Asian countries. Typical among the paste fishery products are '**Bangoong**' of Philippines, '**Ngapi**' of Myanmar, '**Belacan**' of Malaysia and '**Trassi**' of Indonesia. In the North eastern part of India, different fishery products are prepared by the people of different community by their traditional methods. Indigenous fishes seasonally available in large quantities are fermented in Meghalaya, Mizoram, Manipur and Assam. To name, '**Tungtap**' of Meghalaya, '**Namsing**', **'Karoti'** and **'Bardia'** of Assam, '**Ngari**' and '**Hentak**' of Manipur etc. are the few examples of fermented fish from the North East region of India.

In many part of the world, artisanal fish processing remains the predominant and most important method of fish preservation. The methods used mainly are smoking, sun-drying, cooling and freezing, salting, fermentation and grilling and frying. These processes, however, are used alone or in combination so as to achieve the desired product. Curing is adopted by most developing and underdeveloped countries for fish preservation. Curing involves techniques like dry-salting/brining or smoking. Physical dehydration (drying, freeze drying and concentration) play an important role in controlling the rate of deterioration of food. In drying, heat of sun and movement of air remove moisture which causes to dry. Traditionally, whole small or split large fish are spread under the sun on the ground, or on mats, nets, and roofs or on raised racks. The introduction of raised drying racks to artisanal fish curing communities is one of the most widespread research projects in traditional fish processing.

However, the choice of a particular processing method is greatly influenced by the area's geographical location, socio-economic factors and the habits of the local people. The predominant processing method in a country depends upon socio-cultural factors, food habits and the availability of salt. In coastal countries with readily available or cheaper sources of salt, fermented products are usually heavily salted. In Ghana, due to availability of fuel wood and solar salt, a lot of fish is smoked, fermented or salted and dried. In Africa salting and drying of fish for preservation is accompanied by fermentation, but the period is short (a few days) and the product is not transformed into paste or sauce. The products are characterized by a strong odour and, for this reason; various authors have described the product as "sink" fish. In Ghana, fermented fish is called '**momone**', an 'Akan' word which literally means stinking. The "stink" fish of Sierra Leone has been described (Watts, 1965) as fish which had developed a strong odour within 24 hours of capture and was salted for about four days and then dried. Fish fermentation in the Southeast Asian sub-region normally lasts for several months (three to nine months) and the fish flesh may liquefy or turn into a paste (Huss and Valdimarson, 1990). Some of these products include '**nuoc-mam**' of Vietnam and Cambodia, '**nam-pla**' of Thailand, '**sushi**' of Japan and '**patis**' of Philippines. '**plaa-som**' is a Thai fermented fish product prepared from snakehead fish, salt, palm syrup and sometimes roasted rice and the fermentation process last for 5-8 days. Another fermented fish product of Thailand is '**som-fak**' prepared by mixing chopped fish, salt, cooked rice, sucrose, black pepper, water and garlic. The mixture is then packed in banana leaves and fermented for 2-5 days at 30 $^0$C (Christine Paludan-Muller, et al, 2002). "**Pla-ra**' is a prominent lactic acid fermented product of Thailand. During its preparation the fish is scaled, eviscerated, and mixed with salt in the ratio of 3:1 and is then packed in jars, kept for a period extending from 15-90 days. After a stipulated storage period, the fish is taken out of the jars, wash well and allowed to drain. It is subsequently mixed with ground, roasted rice in the ration 1:10 rice to fish and again placed in the jars for further storage for 1-6 months (Reilly, *et al.,* 1990). Unlike Southeast Asian products, fermented fishery products in Africa usually remain whole and firm after processing. The fermentation period is comparatively short, hence the fish muscle does not break down, In some cases whole species are cured and the product retains most, if not all, of its original features so that the species can easily identified.

Fish sauces are popular products in Southeast Asia, where, they are known by various names such as '**Ngapi**' (Burma), '**Nuoc-mam**' (Cambodia and Vietnam), '**Nam-pla**' (Laos and Thailand), '**Ketjap-ikan**' (Indonesia), and so on. The production of some of these begins with addition of salt to uneviscerted fish at a ratio of approximately 1:3, salt to fish. The salted fish are then transferred to fermentation tanks generally constructed of concrete and built into the ground or placed in earthenware pots and buried in the ground. The tanks or pots are filled and sealed off for at least 6 months to allow the fish to liquefy. The liquid is collected, filtered, and transferred to earthenware containers and ripened in the sun for 1 to 3 months. The finished product is described as being clear dark-brown in colour with a distinct aroma and flavor (Mc Kercher et al, 1978).

In North East India, a variety of techniques are employed for preservation of fish. For instance, fish fermentation in Assam is usually carried out in a bamboo cylinder and in earthen pots, in Meghalaya it is done in earthen pots that are kept over ground for fermentation to take place and last for about 3-7 months whereas in Manipur, the fermentation is carried out in earthen pots which are kept underground during the course of fermentation. Sarojnalini and Viswanath (1987) reported that 'Hentak' and 'Ngari' are the two fermented fish products widely consumed in Manipur. Hentak is ball-like thick a paste prepared by fermentation of mixture of sun-dried fish (*Esomus danricus*) powder and petioles of aroid plants (*Alocasia macrorhiza*). Dry fish is crushed to powder, an equal amount of petioles of aroid plants is mixed and a ball-like thick paste is made. The mixture is kept in an earthen pot and is fermented for 7-9 days. Hentak is consumed as curry as well as a condiment with boiled rice. It is given in small amounts to mothers in confinement and patients in convalescence. Ngari is another fish product likable to the Manipuri's because of its special flavour. It is a compulsory item in curry preparations in Manipur. During its preparation, the fish (*Puntius sophore*) is rubbed with salt, sun-dried for 3-4 days, pressed tightly in an earthen pot, sealed airtight and then stored at room temperature for 4-6 months. They also reported that small size fishes e.g. *Puntius ticto, Puntius phutunio, Puntius sophore, Esomus danricus,* etc. which are not readily acceptable for consumption are often sun-dried. They may be converted to a form acceptable to consumer by solid substrate fermentation. During preparation of Ngari, the fish (*Puntius sophore*) is rubbed with salt, dried in the sun for 3-4 days, pressed tightly in an earthen pot, sealed airtight and then stored at room temperature for 4-6 months. Ngari is eaten as a side-dish with cooked rice. Hentak on the other hand is prepared by fermentation of a mixture of sun-dried fish (*Esomus danriecus*) powder and petioles of aroid plants (*Alocasia macrorhiza*). Dry fish is crushed to powder, equal amount petioles of aroid plants are mixed and a ball-like thick paste is made. The mixture is kept in an earthen pot and is fermented for 7-9 days. Hentak is consumed as curry as well as condiment with boiled rice (Thapa 2002).

To prepare Tung tap (*Puntius sarana*), at first the partially cured fishes are used as raw material and the process follows for fermentation is shown in the following flow-chart.

Partially Cured Fish

↓

Washed properly with water,

↓

Salt is applied thoroughly to the fishes (without degutting and scaling),

↓

To the clay pot that can accommodate 5-6 Kg of fishes and which was used earlier for fermentation, fish oil is applied on its outer and inner walls so as to facilitate anaerobic environment during course of fermentation,

↓

The fish is then placed in the clay pot upto its neck,

↓

The mouth of the pot is then filled with mixture of salt and either fish fat or pork fat, covered with banana leaf, and bound tightly at the rim using a jute cord,

↓

The clay pot is then left as such for a period of 3-7 months at room temperature (26±2 $^{0}$C),

↓

After this period, the fish is taken out by removing the extra salt and fat and sold in the market.

(Baishya, 2007)

Preparation of Tungtap in large clay pot

People from various parts of Assam fermented fish and consumed as their regular diet. The small bony fishes (*Puntius ticto*, *P. sophore*, etc.) are selected for the preparation of the product, and locally known as 'sindal'. The people used three types of preparations. In the first type of preparation, the fresh fish are degutted, pounded and mixed with salt; the mixture is packed tightly inside a bamboo cylinder and kept for about 3 months. The second type involved sun drying of the degutted fish, with or without salting, and then tightly packed in an earthen pot for 2-3 months. In the third type of preparation, the fresh fish are degutted, washed, mixed with salt and chilly, and smoked over an earthen-oven for about a month. The 'sindal' prepared in a bamboo cylinder are kept for 6-7 months by keeping the cylinder in inverted position under the water flowing from hill in small stream. This process, although practiced traditionally, the scientific influence behind the technique is to preserve the keeping quality of the finished product at lower temperature.

People of the Eastern Himalayan regions of Nepal, India and Bhutan consume varieties of traditionally processed smoked, sun-dried, fermented and salted fish products (Tamang, 2001). "**Sukako maacha**' and '**Gunchi**' are typical smoked and dried fish products prepared and consumed in eastern Nepal, the Darjeeling hills and Sikkim in India by Nepalis and the Lepcha, respectively. Two types of fishes are preferred for the preparation of 'sukako maacha' by the people residing near streams of river. These hill river fish are mostly 'dothay asala' (*Schizothorax richardsoni*) and 'chuchay asala' (*Schizothorax progastus*) (Tamang, 1992). The fish are collected in a bamboo basket from the river, are degutted, washed, and mixed with salt and turmeric powder. Degutted fish are hooked in a bamboo-made string and are hung above the earthen-oven in a kitchen for 7-10 days. Sukako maacha is kept inside the bamboo-made basket, locally called 'perungo' and can be stored for 3-4 months at room temperature.

'Gunchi' is a typical smoked and dried fish product common to the Lepcha. Fish are caught early morning and are collected in a bamboo basket tied around the waist of the fisherman while fishing. Fish captured include *S. richardsoni*, *Labeo dero*, *Acrossocheilus* sp., *Channa* sp., etc. Fish are degutted, mixed with salt and turmeric powder and hung one after the other in a bamboo stripe above the earthen-oven and smoked for 7-10 days. Gunchi can be kept at room temperature for 2-3 months. As these products are manufactured by the rural people during the appropriate season, they are regarded as a special dish for them (Thapa, 2002).

'**Sidra**' is a sun-dried fish product consumed by the Nepalis living in the Eastern Himalayan regions of eastern Nepal, the Darjeeling hills and Sikkim in India and Bhutan. During its preparation, the whole fish (*Puntius sarana*) is collected, washed and dried in the sun for 4-7 days. Sidra can be stored at room temperature for 3-4 months for consumption. It is usually consumed as pickle. '**Sukuti**' is also a very popular sun-dried fish product among the Nepalis. During preparation of 'sukuti', fish (*Harpodon nehereus*) is collected, washed, rubbed with salt and dried in the sun for 7-4 days, and can be stored at room temperature for 3-4 months.

**Bakasang** is a traditional fermented fish sauce from Indonesia produced by fermenting small whole sardine (*Sardinella* sp., *Stelophorus* sp.) or the guts of big fish (*Katsuwonus pelamis*). The product is prepared by mixing approximately 1.5 to 3.0 parts salts with 5 parts fish, packed into small bottles, and placed at the kitchen near a fire place (temperature ranging from 30 to 60 $^0$C) and allowed to ferment for about 3 to 6 weeks. The product has become closely integrated with the eating habits of Eastern Indonesian people, especially the Manadonese people (in the North Celebes Island). It is usually used as flavouring agent for many dishes or mixed with red chilies, tomato, red onion, and garlic. **Belachan** is a sauce food made by adding suitable amount of salt (5 ~ 20%) to its principal ingredient—shrimp (2 ~ 3 cm) that were caught in a suburban area of Nelaka, Malaka in western Malaysia, where the mouth of the river and coastline intersect each other. The usual ratio between shrimp and salt widely range from 10:1 to 5:1, depending on individual taste. After being caught, the shrimp are spread on wooden boards to be exposed to the sun and dried for several hours. They remain in the sun until their moisture rate falls to around 50%. Then, salt is added to the shrimp in accordance with the individual's favorite ratio and mixed thoroughly. This mixture is put into wooden barrels or other appropriate containers, and stationary fermentation is employed for approximately 1 week, at natural room temperature. In some areas, some rice powder is added to the mixture. After being suitably fermented, the paste-like mixture is poured into molds, and thereafter solidified and dried in the shade until finished. For this product, containers such as plastic packets are used for packing. **Badu** is another sauce food found in Kelatan in western Malaysia. Its constituent ingredients are marine fries of Stolophorus genera and salt water whose concentration is between 10% and 20%. Although the concentration of salt differs depending on the preferences in each area, the most common ratio between fries and salt water is 1:2. Next, this mixture is transferred into earthen bottles or round concrete tanks and permitted to ferment in the shade. After fermenting for approximately 6 months, with stirring every few days during the fermentation period, the upper part of the solution becomes dark brown, transparent liquid in its final stage. This liquid is then withdrawn, boiled, filtrated and finally bottled into product. Also, tomato juice may be added to give it more flavor and taste. The ingredients of **Chinchaluk** (also found in Western Malaysia) are shrimp (mainly marine shrimp, but limnetic shrimp are used in some areas), rice powder and salt. Initially, the shrimp (the principal ingredient) are washed in tap or natural water and dried in the sun for 2 ~ 3 hours. Thereafter, salt water, whose salt concentration is between 10% and 20%, and suitable amounts of rice powder are added and mixed. The basic ratio of its combination is 5:5:1. Finally, this mixture is poured into bottles or large earthen jars and permitted to ferment for approximately 1 month at natural room temperature. Reputedly the best fish sauce is made on Phu Quoc, an island in the Gulf of Thailand, where exquisite and delicate anchovies, called **Ca com**, are layered, salted, and left to ferment for months in their wooden casks. After 3 months, the juice is tapped and poured back on top of the layered fish. Three months later, the liquid is tapped again—and it is this extraction that is considered the "first pressing" and of the highest quality. It's this sauce that goes on the dining table, while second and third pressings are used in cooking. Interestingly, fermented fish sauce was also a great favorite of the Romans. Apicius cited it over 2,000 years ago in his cookbook, calling it **Garum** and **Liquamen**. The taste for it died out in Europe with the end of the Roman civilization except for natives in that tiny Roman outpost of Great Britain.

## References

Baishya, D. (2007). Formulation of Starter Culture for fermentation of Fish, on the basis of Microbiological, Biochemical and Molecular Analysis. Ph.D Thesis, Gauhati University, Guwahati, India.

Beddows, C.G. (1985). The History of Fermented Foods. In: Handbook of Fermented Functional Foods, ed. Edward R. Farnwarth, CRC press, USA. p 22.

Christine Paludan-Mullar, Mette Madsen, Pairat Sophanodora, Lone Gram and Peter Lange Moller (2002). Fermentation and microflora of *plaa-som*, a Thai fermented fish product prepared with different salt concentrations, *International Journal of Food Microbiology*. 73(1): 61-70.

Gopakumar, K. (1997). Tropical fishery products. Oxford and IBH Publishing Co., New Delhi. p 174.

Huss, H. H. and Valdimarson, G. (1990). Microbiology of salted fish. FAO Food Tech. News, 10 (1): 190.

McKercher, P. D., Hess, W. R. and Hamdy, F. (1978). Residual viruses in pork products. *Appl. Environ. Microbiol.* 35: 142-145.

Reilly, P.J.A., Parry, R.W.H. and Barile, L.E. (1990). Post-harvest technology. Preservation and Quality of dish in South in South East Asia. International Foundation of Science, Stockholm. pp 176-177.

Sarojnalini, C. H. and Viswanath, Singh (1987). Composition and Digestibility of Fermented fish foods of Manipur. *J. ed. Sci. Technol.* 25(6): 349-351.

Tamang, J.P. (1992). Systematics, Distribution and Ecology of the Icthyospecies of Sikkim and their Bearing on the Fish and Fisheries of the State. Ph.D Thesis, Gauhati University Guwahati, India.

Tamang, J.P. (2001). Food culture in the Eastern Himalayas. *J. Himalayan Research and Cultural Foundation.* 5 (3-4): 107-118.

Thapa, N. (2002). Studies of microbial diversity associated with some fish products of the Eastern Himalayas. Ph.D. Thesis, North Bengal University, India.

# Chapter 3

# Microbial Diversity

The traditional fermentation process is an uncontrolled system and is usually carried out in a quite unhygienic environment. As the process of fermentation is brought about by the microbial activity so in the traditional system of fermentation there is every possibility of inclusion of microorganisms which is not desired in or that can cause spoilage to the finished product. Although no incidents of food poisoning has been reported due to consumption of fermented fish there are also no reports of any clinical investigations in this regard. A study on the microflora of such a fermented fish product is useful to understand - how safe is the product for consumption. The diverse groups of microorganisms that are present in the fermented fish products will provide information about their characteristic nature - beneficial or pathogenic, their contribution towards production of flavour and aroma, etc. The environment, however, is the key factor for the so called 'microbial diversity' among different fermented fish products round the globe. Moreover, the aquatic habitat, processing technique and of course, the handling during processing are the other factors that equally make a way for the introduction of a variety of microorganisms in the finished product. A study on different curing stages i.e. fresh, semi-cured and fermented, is therefore necessary to understand the pattern of microbes and become evident the role played by the microbes during fermentation.

**Deterioration Process in fish**

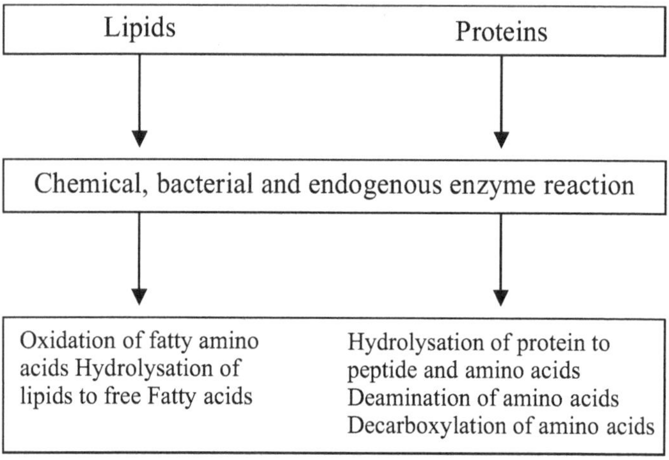


Fish in its natural environment has its own micro-flora in the slime on its body, in its gut and in its gills. These micro-organisms, as well as the enzymes in the tissues of the fish, bring about putrefactive changes in fish when it dies. Furthermore, the microorganisms generally present in the salt used for salting also contributes to the degradative changes in fish. The bacterial genera occurring in fresh fish in order of abundance are *Achromobacter, Vibrio, Pseudomonas, Flavobacterium* and *Micrococci*. The bacterial loads in different part of a freshly caught marine fish are of the following order: skin $10^3$-$10^5/cm^2$; gills $10^5$-$10^6/cm^2$; intestine $10^5$-$10^8/cm^2$ (Chichcester and Graham, 1973). The numbers of microorganisms on the skin of fish can be influenced by the method of catching. For example, trawling fish nets along the bottom for long period results in exposure of fish to high bacterial counts in the disturbed bottom sediment, and this can be reflected in the initial microbial load on the fish (William *et* al., 1995). The flora of fresh-caught fish from northern waters are usually gram negative rods which include *Pseudomonas* and *Moraxella* type species (*Achromobacter* or *Acinobacter*) with some *Coryneforms, Flavonbacterium, Micrococcus* and *Cytophaga* species also being present (Nickerson and Sinskey, 1972). It has also been stated that among the different species of *Pseudomonas*, 40-50 % are *P. pellucidium*, 20-30 % *P. geniculatum*, 10-20 % *P. schuylkillensis* and *P. fluorescence.* Species related to *P. ovalis, P. fragi* and *P. multistriatum* are also present in small numbers. It would appear that fish from warmer waters such as the Indian Ocean or the African coastal waters, Coryneforms and Micrococcus may comprise the major constituents of the flora of certain fish when removed from the water. Regardless of the flora originally present on the fish removed from water, it is considered that freshwater bacterial types will take over and eventually become the predominating flora due to the manner in which fish are handled (Nickerson and Sinskey, 1972). A study on fresh, partially cured and fermented *Puntius sarana* showed a maximum recorded bacterial load of $1.92 \times 10^6$ c.f.u/ml followed by $1.66 \times 10^6$ c.f.u/ml from skin of fermented fish and partially cured fish respectively while the lowest was $1.02 \times 10^6$ c.f.u/ml followed by $1.06 \times 10^6$ c.f.u/ml from the intestine of fresh and partially cured fish respectively. The result showed that there is a wide variation in the bacterial load in the samples at different states. One of the reasons may be attributed to the fact that skin is exposed to the outside environment (air and water) or the microbial content of the aquatic habitat of the samples also attributed for the variation recorded. High microbial load may also be due to improper handling and unhygienic keeping quality during the curing process. Transportation stresses might have also contributed to the high microbial load. Quantitative investigations of bacterial analysis of the market samples of iced and salted hilsa (*Hilsa ilishsa*) revealed the highest bacterial load in the intestine of iced hilsa followed by gill tissue and lowest count was found on the skin surface. Qualitative analysis indicated that *Micrococcus* species as the most predominant in all parts of iced and salted hilsa. Other bacteria were identified as *Staphylococcus* sp., *Pseudomonas* sp., *Bacillus* sp. and *Achromobacter* sp. (Hog *et al.*, 1983). Table 1 projects various microbial load of a few fermented fish species from eastern Himalayan region of India.

**Table 1.** Total microbial load (c.f.u/g of sample) of different fermented fishes

| Sample | LAB | Endospore forming bacteria | Yeast | Aerobic mesophilic count |
|---|---|---|---|---|
| Tungtap | 6.2 | 3.0 | 2.4 | 6.8 |
| Nagri | 6.7 | 4.2 | 3.0 | 7.0 |
| Hentak | 4.6 | 3.8 | <1 | 4.2 |

Microorganisms require water in an available form for growth and metabolism. It is found that at water activity ($A_w$) below 0.6; all the microbial growth is inhibited. The water activity ($A_w$) of traditionally fermented Tung tap is found to be 0.798 indicating an environment supportive of microbial growth. While halophiles grow at high salt concentration, are unable to grow in salt free media. Halotolarent organisms grow best without significant amounts of salt but can also grow in concentrations higher than that of sea water. Various types of salts are used for salting and fermentation of fish. They include solar salt, rock salt, and vacuum salt and have their own micro-flora. Solar salt, which is the most widely used in fish curing, has been found to contain the largest amount of micro-organisms. The general bacterial flora of solar salt mostly comprises Bacillus types (75 %) with the remainder being Micrococcus and Sarcina types. The most important spoilage organisms always present in solar salt are the red halophilic bacteria. The red halophiles belong to two genera of bacteria, namely Halobacterium and Holococcus. Halobacterium consists of rod-shaped bacteria and requires at least 10-15 % salt concentration whilst Holococcus can thrive at 5-10 % salt concentration (Kreig and Holt, 1984).

A study by Knochel and Huss (1984) on microbiology of barrel salted herrings revealed that both aerobic and anaerobic viable counts (in media containing 15 percent sodium chloride) were found not more than $3x10^5$/g of fish. The types of microorganisms identified were- gram positive aerobic Halotolarent cocci (20 %); gram negative aerobic halophilic rods (70 %) and yeasts (3 %).

Studies on the microbiological changes in 'Bagoong', a fermented fishery product of Philippines, showed that the total viable count decreased with time. Aerobic organisms predominate at the onset of fermentation followed by microaerophilllic and anaerobic microorganisms at the later stage. The micro-flora indicated that both desirable and hazardous rapidly up to the sixth month and declined slightly until the end of fermentation. Most of the organisms isolated were facultative anaerobes (Van ver, 1965).

Microorganisms that are introduced in the product before drying or are introduced during processing can survive for extended periods. This is most important with respect to pathogens if they are present in hazardous numbers before drying or if time and temperature allow them to resume growth in a product that is dehydrated before consumption. There have been a number of instances where the survival of pathogen or their toxins has caused problems in products. Intermediate moisture foods, IMFs, are commonly defined as those foods with a water activity between 0.85 and 0.60. This range, which correspondence roughly to a moisture content of 15-50%, prohibits the growth of gram negative bacteria as well as large number of gram positive bacteria, yeasts and moulds, giving the products an extended shelf life at ambient temperature (Adams and Moss, 2003).

McKercher *et al.* (1978) reported the growth of halophilic aerobic spore formers as predominant microorganisms from fish sauces prepared by fermenting them. Lower numbers of Streptococci, Micrococci and Staphylococci were found, and they, along with the *Bacillus* sp. were apparently involved in development of flavor and aroma.

Fish spoilage is a complex process involving both non-microbiological and microbiological processes. Non-microbiological deterioration is caused by endogenous proteolytic enzymes, which are concentrated in the head and viscera and attack these organs and surrounding tissues after death. Enzymatic spoilage is followed by the growth of microorganisms, which invade the fish flesh, causing breakdown of tissues and a general deterioration of the product. During processing of catfish (e.g., deheading, eviscerating, cutting), the microorganisms present in the surface slime layer, the gills and the gut can be spread onto the processing equipment, the workers and the flesh of the fillet. Hence, the normal sterile flesh can be contaminated with millions of bacteria (Banwart, 1989; Bonnell, 1994; Garthwaite, 1997; Inglish *et al,* 1993).

The bacteria most often involved in the spoilage of fish are part of the natural flora of the external slime of fishes and their intestinal content. There are three main routes through which bacteria penetrate from the outer and inner surface into the fish flesh:

- Gills to blood vessels to flesh;
- Intestine to body walls to flesh and
- Skin (slime) to flesh. (Doyle & Decker, 1989)

Once the fish dies, its natural defense mechanism fails and bacteria become active, which secrete digestive enzymes. Bacteria and enzymes break down and dissolve the tissue they attack. Flesh of fish contains considerable amounts of non-protein nitrogen compounds. The enzymes naturally present in the flesh produce autolytic changes that increase the supply of nitrogenous food like amines and amino acids and glucose for bacterial growth. The bacteria convert these compounds to their break down products, the end product being hydrogen sulphide, mercaptanes, indole, etc. that are indicative of putrefaction (Doyle & Decker, 1989).

The spoilage flora of iced northern fish consists almost entirely of *Pseudomonas* and *Achromobacter* sp. And usually 20-50 percent of the *Pseudomonas* is *P. fragi* type. Volatile bases are produced during the growth of spoilage bacteria in fish, but trimethylamine is not necessarily a significant component of such bases when spoilage occurs. The *Achromobacter* sp., though present as a part of the spoilage flora, would appear to play a minor role in the production of organoleptic changes (Roberts and Skinner, 1983).

Alur *et al.* (1971) isolated *Pseudomonas, Proteus, Aeromonas* and *Achromobacter* from spoiled fish. *Pseudomonas* and *Proteus* cause putrid and ammoniacal odors, while *Acinobacter* and *Aeromonas* are associated with unpleasant sweetish or fruity odors.

Investigations on Cod and Haddock have indicated that *Pseudomonas putrefaciens* (Produce $H_2S$ from proteins) is especially active in causing spoilage of whitefish form the North Atlantic (Tarr, 1954).

A recent study on Haddock (Reilly et al, 1990) showed that, in fish iced two days (fresh), *Achromobacter* sp. was the predominating flora, some *Pseudomonas* and *Flavobacterium* species also being present. As the fish was held (Five days, borderline quality) the *Achromobacter* further increased, the proportion of *Pseudomonas* remaining approximately constant while *Flavobacterium* species dropped out.

Once it is considered that the spoilage of fresh fish is entirely a surface phenomenon, in that the bacteria grow on skin and body cavity lining surfaces and their products diffuse into the tissues to cause organoleptic changes. On the other hand, at the time of advanced spoilage flesh counts as high as $10^7$-$10^8$ per gram may be obtained and there may be obvious invasion of bacteria into the tissue, through the skin and even through the bacterial flora of fresh water or marine water. (Nickerson and Sinskey, 1972).

William *et al.* (1995) indicated that discolorations of the fish flesh may occur during spoilage; yellow to greenish-yellow colours caused by *Pseudomonas fluorescens*, yellow by Micrococci, and others; red to pink colours from the growth of *Sarcina, Micrococcus*, or *Bacillus* sp.

As with meat, the muscle and internal organs of healthy, freshly caught fish are usually sterile but the skin; gills and alimentary tract all carry substantial numbers of bacteria. Their numbers as reported, on skin ranged from $10^2$ -$10^7$ c.f.u/$cm^2$ and from $10^3$- $10^7$ c.f.u/g in the gills and the gut. These are mainly gram negative bacteria of the genera *Pseudomonas, Salmonella, Psychrobacter, Vibrio,*

*Flavobacterium* and *Cytophaga* and some gram positive bacteria such as *Coryneformis* and *Micrococci*. From the fresh, partially cured and fermented *Puntius sarana*, a total of 14, 23 and 20 numbers of bacterial isolates were recovered respectively. Out of 14 bacterial isolates in case of fresh fish sample and 23 bacterial isolates from partially cured fish, 21.43% and 17.39% each were from gill and intestine respectively while 57.14% and 65.22% was recorded from skin respectively. On the other hand, out of 20 bacterial species isolated in case of fermented fish, 40% each were from intestine and skin while 20% was from gill (Table 2).

**Table 2.** Total number and occurrence of bacterial isolates from different parts of *Puntius sarana*.

| Sample Type | Total | No. of Isolates in Gill | Percentage of Occurrence (%) | No. of isolates in Intestine | Percentage of Occurrence (%) | No. of Isolates in Skin | Percentage of Occurrence (%) |
|---|---|---|---|---|---|---|---|
| Fresh | 14 | 3 | 21.43 | 3 | 21.43 | 8 | 57.14 |
| Partially Cured | 23 | 4 | 17.39 | 4 | 17.39 | 15 | 65.22 |
| Fermented | 20 | 4 | 20.00 | 8 | 40.00 | 8 | 40.00 |

A total of 67 microbial strains were isolated from 'momoni', a Ghanaian fermented fish condiment obtained from retail outlets. The strains belonged to nine genera of microorganisms namely *Bacillus, Lactobacillus, Pseudomonas, Pediococcus, Staphylococcus, Klebsiella, Debaryomyces, Hansenula* and *Aspergillus* with *Bacillus* having a predominant occurrence of 37.7 per cent (Sanni et al, 2002). Table 3, 4 and 5 shows the different bacterial isolates identified from various parts of fresh, partially cured and fermented *Puntius sarana*.

**Table 3.** Identified bacterial Isolates from gill, intestine and skin of fresh *Puntius sarana*.

| Gill | Intestine | Skin |
|---|---|---|
| (FG1) *Lactobacillus* sp. | (FI4) *Planococcus* sp. | (FS7) *Micrococcus* sp. |
| (FG2) *Leuconostoc* sp. | (FI5) *Bacillus subtilis* | (FS8) *Lactobacillus.* sp. |
| (FG3) *Planococcus* sp. | (FI6) *Sporolactobacillus* sp. | (FS9) *Streptococcus* sp. |
| | | (FS10) *Planococcus* sp. |
| | | (FS11)*Lactobacillus coryneformis* |
| | | (FS12) *Lactobacillus. plantarum* |
| | | (FS13) *Micrococcus luteus* |
| | | (FS14) *Staphylococcus* sp. |

*Lactobacillus* sp. were predominant in case of fresh, partially cured and fermented fish samples. The higher number *Lactobacillus* sp in all parts of partially cured fish may be due to initiation of the fermentation in the sample. Thapa et al. (2004) reported Lactobacilli as the most common isolate from traditional cured fish. *Lactobacillus plantarum* from Ngari; and *lactobacillus fructosus* and *Lactobacillus amylophilus* from Hentak has also been isolated by Thapa et al. (2004). *Lactobacillus* species were also reported from other Asian fermented fish products such as 'nam-pla' and 'kapi'- the fermented Thai fish (Tanasupawat and Komagata, 1992). Out of eight numbers of Lactobacillus species identified from different parts of the partially cured fish, *Lactobacillus fermentum* was encountered from gill, intestine and skin suggesting its possible role in transformation of skin, intestine and gill components by aerobic and anaerobic metabolic processes. The other bacterial species isolated were *Bacillus* sp., *Leuconostoc* sp, *Staphylococcus* sp, *Streptococcus* sp, *Sporolactobacillus* sp, *Corynibacterium* sp, *Klebsiella* sp, *Micrococcus* sp, *Sporosarcina* sp and *Planococcus* sp. Sanni et al. (2002) also reported the strains belonged to nine genera of microorganisms namely endospore forming *Bacillus* sp., *Pediococcus* sp., *Staphylococcus* sp., *Pseudomonas* sp., *Klebsiella* sp., *Lactobacillus* sp., etc. in the fish products.

Two species of Micrococcus were identified from the skin of fresh, partially cured and fermented fish samples. Presence of these organisms suggests a possible introduction of these organisms from the environment during storage. *Micrococcus* sp. was also reported from some fermented fish products of Thailand (Phithakpol, 1993) and Japan (Wu *et al.,* 2000).

Out of the Bacillus species isolated from different parts of fresh, partially cured and fermented fish, *Bacillus subtilis* was found to occur in the intestine of the fresh and fermented fish and not found from gill or skin. The reason may be attributed to anaerobic nature of the organism. However, the other *Bacillus* sp. isolated from partially cured fish were encountered from skin suggesting the possible facultative anaerobic nature among the species. *Bacillus* sp. was also found to be reported in fish product and the reason might be due to their ability to form endospore to survive under the prevailing conditions (Crisam and Sands, 1975). Although three *Bacillus* sp. were found in the partially cured fish sample, in case of fermented fish only one isolate could be recovered. This could be the impact of competition and/ or antagonistic effect of the predominant lactic acid bacteria that have prevented the proliferation of the *Bacillus* sp. which was also observed by Adams and Nicolaides (1997).

There is very little information on Salmonella food poisoning arising from the consumption of fermented fish in Africa despite the unsanitary fish processing practices observed in many countries. In a study conducted by Nerquaye-Tetteh *et al.* (1978) to isolate various micro-organisms, no *Salmonella* sp. were found from samples of fermented fishery product obtained from the open markets in Ghana. Either from partially cured or fermented *Puntius sarana* no *Salmonella* sp. and *E. coli* could be isolated. However, it has been stated that the recovery of the organisms also depends on the efficacy of the sampling procedure, size of the sample and the method of examination including choice of media and incubation temperature (Roberts, 1982).

**Table 4.** Identified bacterial isolates from gill, intestine and skin of partially Cured *Puntius sarana*.

| Gill | Intestine | Skin |
|---|---|---|
| (BG17)*Staphylococcus aureus* | (BI23)*Klebsiella ozaenae* | (BDS1)*Lactobacillus fermentum* |
| (BG18) *Streptococcus faecalis* | (BI25)*Klebsiella pneumoniae* | (BDS2)*Lactobacillus ruminis* |
| (BG19) *Sporosarcina* sp. | (BI26)*Lactobacillus fermentum* | (BDS3)*Bacillus* sp. |
| (BG20)*Lactobacillus fermentum* | (BI27) *Planococcus* sp. | (BDS4)*Leuconostoc dextranicum* |
| | | (BDS8)*Lactobacillus casei* |
| | | (BDS29)*Micrococcus varians* |
| | | (BDS30) *Corynebacterium xerosis* |
| | | (BDS31)*Planococcus* sp. |
| | | (BVS13)*Bacillus polymyxa* |
| | | (BVS14)*Lactobacillus lactis* |
| | | (BVS35)*Lactobacillus casei* |
| | | (BVS36)*Corynebaterium kutscheri* |
| | | (BVS38)*Bacillus megaterium* |
| | | (BVS39)*Micrococcus. roseus* |
| | | (BVS40)*Lactobacillus fermentum* |

**Table 5.** Identified bacterial isolates from gill, intestine and skin of fermented *Puntius sarana.*

| Gill | Intestine | Skin |
|---|---|---|
| (FFG1) *Lactobacillus* sp. | (FFI7)*Sporosarcina* sp. | (FFS3)*Micrococcus* sp. |
| (FFG2)*Leuconostoc* sp. | (FFI8)*Streptococcus* sp. | (FFS4)*Lactobacillus* sp. |
| (FFG5)*Planococcus* sp. | (FFI9)*Staphylococcus* sp. | (FFS15)*Streptococcus* sp. |
| (FFG6)*Staphylococcus* sp. | (FFI0)*Planococcus* sp. | (FFS16)*Planococcus* sp. |
| | (FFI11)*Bacillus subtilis* | (FFS17)*Lactobacillus coryneformis* |
| | (FFI12)*Sporolactobacillus* sp. | (FFS18)*Lactobacillus Plantarum* |
| | (FFI13)*Micrococcus varians* | (FFS19)*Micrococcus luteus* |
| | (FFI14)*Planococcus* sp. | (FFS20)*Staphylococcus* sp. |

Moulds are able to grow under dry condition better than bacteria. For this reason, moulds are often associated with dried fishery products. Spores of moulds which are often present in the air and soil contaminate fish during processing. The moulds commonly associated with dried cured fish in storage are *Aspergillus halophilus, A. restrictus, Wallemia sebi, A. glaucus, A. candidus, A. ochraceus, A. flavus* and *Penicillum* sp. (Christensen and Kaufmann, 1974).It is reported that all mould growth is influenced by temperature. The optimum temperature for mould growth is 30$^0$C and the maximum ranges from 40$^0$ to 55$^0$C depending on the species. Reported incidents of aflatoxin poisoning from fish products in Africa are rare, probably due to the long period of heating during food preparations or lack of records. In case of partially cured *Puntius sarana* and traditionally fermented tungtap the fungal count was found to be 16 c.f.u/plate and 17 c.f.u/plate respectively. The fungal species identified from partially cured fish were

*Oideodendron* sp., *Fusarium* sp., *Aspergillus nidulans* and *Rhizopus* sp. while that from tungtap were *Fusarium* sp., *Aspergillus nidulans* and *Rhizopus* sp. (Fig 1 and Fig 2).

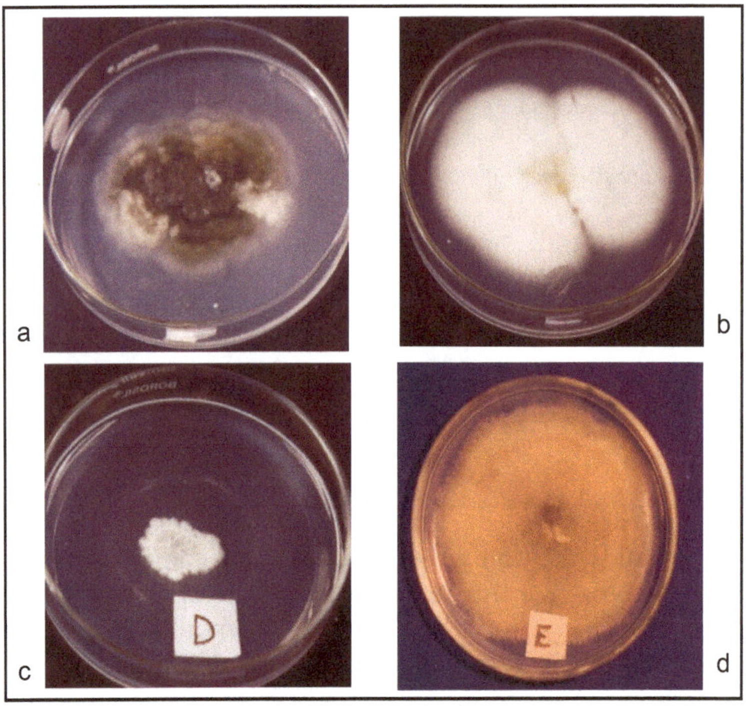


**Fig1**: Different fungal species from Partially Cured *Puntius sarana*, a) *Oideodendron* sp. b) *Fusarium* sp. c) *Aspergillus nidulans* and d) *Rhizopus* sp.

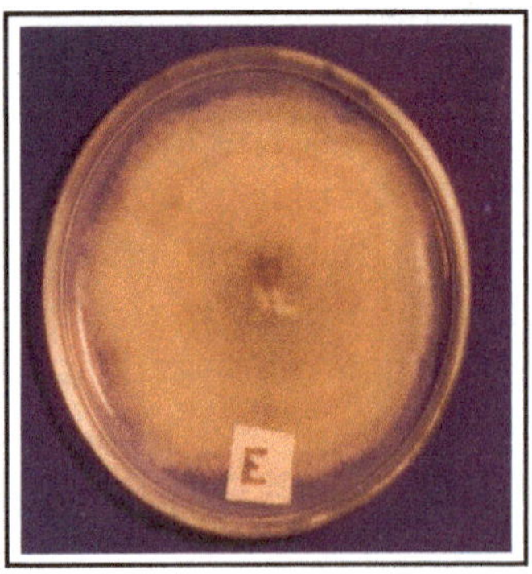


**Fig 2**: *Aspergillus* sp identified from fermented fish, Tungtap

Okonkwo *et al.* (1977) reported dangerous level of aflatoxin in dried fish. Townsend et al. (1971) isolated *A. flavus* from Vietnamese cured fish and identified it as responsible for producing aflatoxin.

A type of spoilage known as 'dun' may occur in salt dry fish especially in light or medium cures, small black, brown, or fown-coloured spots develop on the fish surfaces in this case. This type of spoilage is due to the growth of moulds of the genera *Sporendonema* and *Oospora*. These organisms require 5-10 % of sodium chloride in the medium for growth and some strains can tolerate 20 % of sodium chloride. The source of these organisms is the salt used in curing (Nickerson and Sinskey, 1972).

Microorganisms produce intracellular and extracellular enzymes. Extracellular enzymes (carbohydrases and proteases) are easier to obtain than intracellular enzymes (glycolytic and tricarboxylic acid cycle enzymes). There are cases in which enzymes can be used more effectively than live microbial cells (Banwart, 1987).

Fish being rich in proteins and lipids are susceptible to microbes utilizing these as energy sources by production of protease and lipase. Many of them can be industrially useful. Also enzymes including protease play a major role in fermentation in which chemical changes are brought about in an organic substrate through the action of enzymes elaborated by microorganisms. However, it has been reported that the proteolytic activity, which is caused mainly by tissue proteases (e.g. cathepsins) and to a lower degree, by gut proteases, it is doubtful whether fish fermentation can be utilized on an industrial scale without improving the technique including commercially acceptable inhibitors of proteolytic activity (Fransen, 1983). Lipids in fish have a high content of polysatuarted fatty acids. This caused oxidative rancidity brought about by the microbial flora of the fish during the process of curing. Hydrolysis by lipases results in fatty acids and glycerol or other alcohols. Fat subjected to this type of changes may contain fatty oxy- and hydroxy acids, glycerol and other alcohols, aldehydes, ketones, and lactones; in presence of lecithin, they may include trimethylamine, with its fishy odour (Frazier and Westhoff, 2005). The off-flavour of the finished product can also be attributed to oxidative rancidity caused by secretion of lipase enzyme by the bacterial members associated with the fermentation process. The role of amylase producing members during fermentation in hydrolyzing the starch components of the product can also not be ruled out. Microorganisms involved in the fermentation of 'Burong bangus', some isolates are found to be capable of hydrolyzing starch (Spencer and Hughes, 1963).

The bacterial members isolated from partially cured *Puntius sarana* when screened for their lipolytic, proteolytic and amylolytic activity revealed low (*Lactobacillus ruminis, Bacillus* sp., *L. casei, L. fermentum, L. lactis*, etc.) to moderately higher (*Micrococcus roseus, Sporosarcina* sp.) lipolytic activity (Fig. 3) and in some cases no lipolytic activity (*Micrococcus varians, Bacillus polymyxa*) is seen. The proteolytic activity, on the other hand was found to be highest in case of *Lactobacillus ruminis* while in other bacterial species the activity was low (*Planococcus* sp., *Lactobacillus lactis, L. fermentum*) and moderately higher (*Bacillus* sp., *Lactobacillus casei, Bacillus megaterium*, etc.), (Fig. 4). Bacterial members like *Micrococcus varians, Staphylococcus aureus, Streptococcus faecalis* and *Sporosarcina* sp. showed no proteolytic activity. Highest amylolytic activity was shown by *Bacillus polymyxa* while *L. casei, L. fermentum, L. lactis, Micrococcus varians* etc. have shown moderately higher enzyme activity (Fig. 5) and low amylolytic activity is recorded in case of *Leuconostoc dextranicum, Corynebacterium xerosis, Staphylococcus* sp., *Planococcus* sp. etc. Some bacterial species are found without any amylolytic enzyme activity (*Corynebacterium kutscheri, Bacillus megaterium, Klebsiella ozaneae*, etc.).

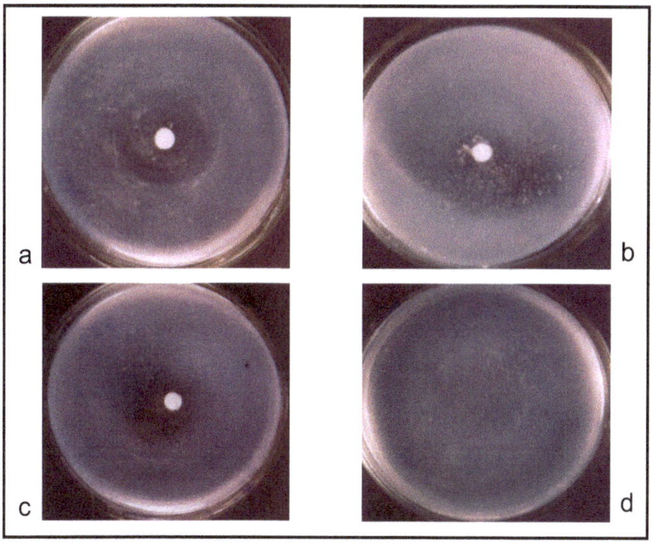

**Fig 3**: Lipolytic activity shown by bacterial isolates from partially cured *Puntius sarana* on tributyrine-agar medium; a) *Lactobacillus lactis*, b) *Lactobacillus fermentum*, c) *Sporosarcina* sp. and d) Control.

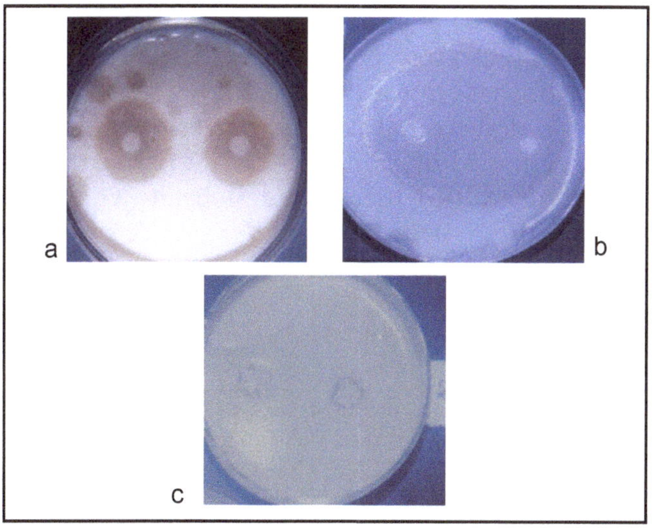

**Fig 4**: Proteolytic activity shown by different bacterial isolates from partially cured *Puntius sarana* on skim-milk agar medium; a) *Lactobacillus ruminis*, b) *Micrococcus cryophylus* and c) *Lactobacillus fermentum*.

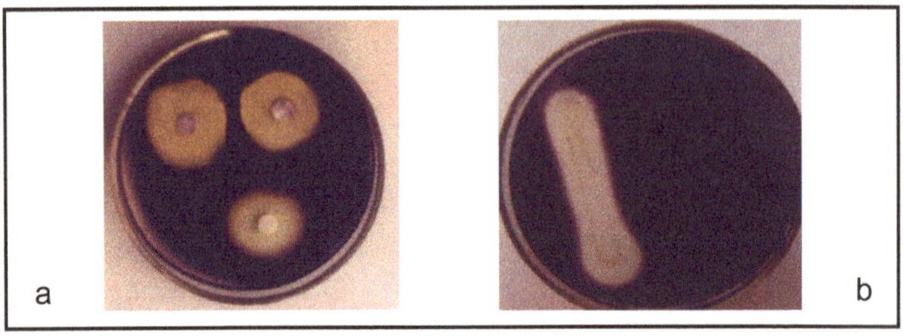


**Fig 5**: Amylolytic activity shown by the two different bacterial isolates from partially cured *Puntius sarana* on starch-agar medium; a) *Lactobacillus fermentum* and b) *Lactobacillus lactis*.

As stated earlier, *Lactobacillus* sp. were predominant in case of fresh, partially cured and fermented fish (Table 3, 4 and 5). Again, one of the remarkable observations was the absence of enteric bacteria such as *Salmonella* sp., *E. coli* or lactose fermenting *Klebsiella* sp in the fermented fish sample, although it was isolated from the partially cured stage. The reason may be the antagonistic effect of the microbial members or selective inhibitory effect of the fish components. It can also be assumed that the fish samples might contain some antimicrobial agents and that the growth of these pathogenic bacteria was restricted (Table 6). The test of antibiosis of the fish extract from partially cured and fermented fish showed zone of inhibition of *E. coli*, *Salmonella* sp., *Klebsiella* sp., *Staphylococcus* sp. and *Streptococcus* sp. (Fig 6 and Fig 7).

**Table 6.** Antimicrobial activity of partially cured and fermented *Puntius sarana* extracts.

| Pathogenic species | Zone of Growth Inhibition (in mm) | | | |
|---|---|---|---|---|
| | Partially Cured Fish Extract | | Fermented Fish Extract | |
| | After 24 hours | After 48 hr | After 24 hr | After 48 hr |
| *E. coli* | 3 | 5 | 2 | 5 |
| *Salmonella* sp. | 5 | 5 | 5 | 5 |
| *Klebsiella* sp. | 4 | 5 | 2 | 3 |
| *Staphylococcus* sp. | 3 | 4 | 3 | 4 |
| *Streptococcus* sp. | 3 | 3 | 3 | 3 |

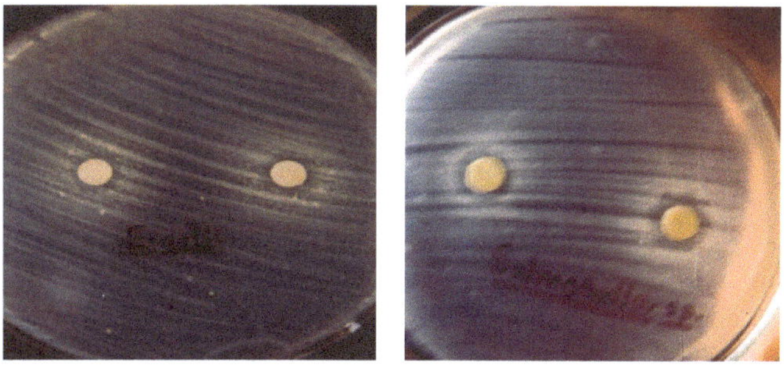


**Fig 6**: Zone of inhibition shown by methanol extract of partially cured *Puntius sarana* against a) *E. coli* and b) *Salmonella* sp.

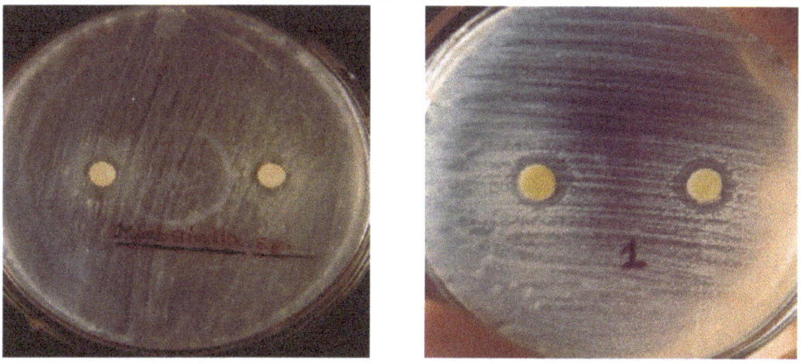


**Fig 7**: Zone of inhibition shown by methanol extract of fermented *Puntius sarana* against a) *Klebsiella* sp. and b) *Staphylococcus* sp.

The microorganisms isolated from partially cured and fermented *Puntius sarana* when cross-inoculated in order to observe their antagonistic activity, the species of *Lactobacillus, Sporosarcina* and *Bacillus megaterium* isolated showed mutual growth while species of *Lactobacillus* showed antagonistic activities against *Staphylococcus* sp. and *Streptococcus* sp. This can be the reason why pathogenic species of *Staphylococcus* and *Streptococcus* were counted in small number in the partially cured and fermented fish samples (*Puntius sarana*). The same reason can be attributed for reduction in the number of undesired microorganisms in the fish products as well as it helps in the preservation of fish as reported by Einarsson and Lauzon, (1995).

The traditional system of fish preservation, which include processes like solar drying or smoking being uncontrolled system, increases the risk of foodborne hazards. A biological hazard is one that, if not properly controlled is reasonably likely to result food borne illness. The biological hazard includes different pathogenic microbes that may render a food unsafe for consumption. In poultry, a simple method of Congo Red Dye binding test (CR test) has been used to differentiate between invasive and non-invasive *E. coli* (Berkhoff and Vinal, 1986; Panigarhy and Yushen, 1990) but its use for detection of other pathogenic bacteria from any fermented food products has not been reported. As the traditional method of fermenting tungtap is not carried out in a controlled environment, therefore, the method like Congo Red Dye binding test was performed to detect the pathogenicity of the organisms isolated from the partially cured and fermented fish samples along with DNase activity test. From the partially cured fish sample two bacterial species viz. *Staphylococcus* and *Streptococcus* have shown positive Congo red binding (Fig 8) and DNase activity test (Fig 9) while from the fermented fish sample three bacterial species viz. *Staphylococcus* sp, *Streptococcus* sp and *Bacillus subtilis* have shown both the tests positive (Fig 10) thereby indicating their pathogenic/toxigenic nature (Table 7). *Bacillus subtilis* is normally not a human pathogen. But in certain cases this organism may show positive DNase agar test possibly to hinder the progression of bacteriophage infection. The percentage of occurrence of *Staphylococcus aureus* and *Streptococcus faecalis* was found to be 3.84%, which is very negligible to cause any health problem after consumption and moreover, the fermented fish item is cooked before consumption. Therefore, from this view the positive DNase activity test and Congo red Dye binding test of these two isolates were found to be insignificant. But, it will be noteworthy to state that in one of the preparation method of this fish product i.e. tungtap is consumed by the producer community as it is along with chilly, salt, mustard oil, onion, garlic etc, and without boiling or cooking. In this respect, the pathogenic strains may cause either a direct or indirect health effect to the consumers. This type of experimentation can, therefore, be used as useful criterion for selecting starter culture for fermenting such fish product under controlled environment.

**Table 7.** Congo red binding test and DNase agar test of bacterial isolates from partially cured and fermented *Puntius sarana*.

| S.No. | Bacterial Species | Congo Red Binding Test | DNase Agar Test |
|---|---|---|---|
| | **From Partially Cured Fish** | | |
| 1 | *Micrococcus varians* | +ve | -ve |
| 2 | *Corynebacterium xerosis* | +ve | -ve |
| 3 | *Bacillus polymyxa* | -ve | +ve |
| 4 | *Corynebacterium kutscheri* | +ve | -ve |
| 5 | *Bacillus megaterium* | +ve | -ve |
| 6 | *Micrococcus roseus* | +ve | -ve |
| 7 | *Klebsiella ozaenae* | +ve | -ve |
| 8 | *Klebsiella pneumoniae* | +ve | -ve |
| 9 | *Staphylococcus aureus* | +ve | +ve |
| 10 | *Streptococcus faecalis* | +ve | +ve |
| 11 | *Sporosarcina* sp. | +ve | -ve |
| 12 | *Lactobacillus fermentum* | -ve | -ve |
| 13 | *Lactobacillus lactis* | -ve | -ve |
| | **From Fermented Fish** | | |
| 14 | *Micrococcus* sp. | +ve | -ve |
| 15 | *Planococcus* sp. | -ve | -ve |
| 16 | *Staphylococcus* sp. | +ve | +ve |
| 17 | *Sporosarcina* sp. | +ve | -ve |
| 18 | *Streptococcus* sp. | +ve | +ve |

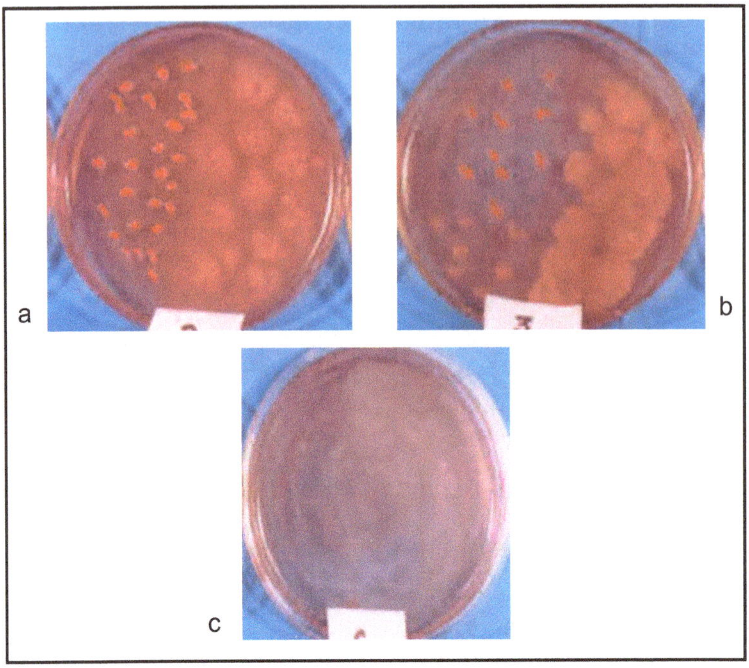

**Fig 8**: Congo Red binding test showing positive results in a) *Staphylococcus aureus*, b) *Streptococcus* sp. isolated from partially cured *Puntius sarana*, c) Control

**Fig 9**: *Staphylococcus aureus* isolated from partially cured *Puntius sarana* showing DNase positive test

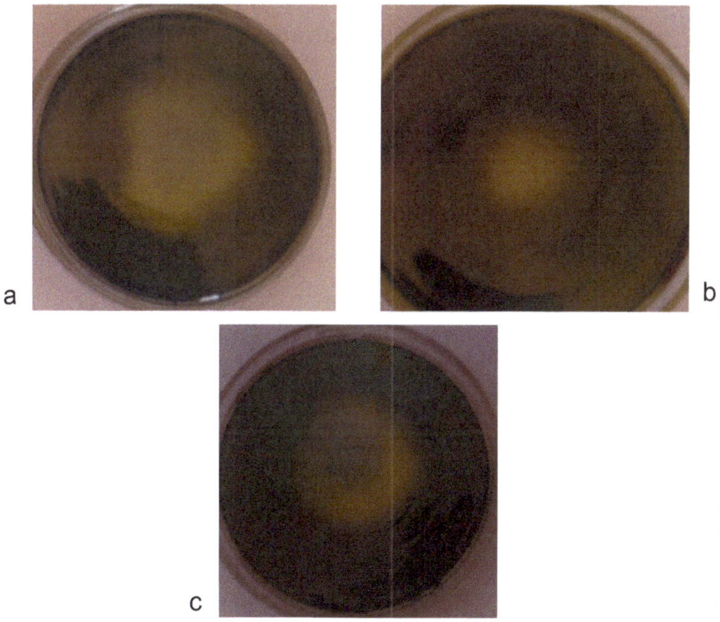


**Fig 10**: Bacterial isolates from Tungtap showing DNase positive test; a) *Bacillus subtilis*, b) *Streptococcus faecalis*. and c) *Staphylococcus aureus*

Two important transaminase enzymes glutamate pyruvate transaminase (GPT) and glutamate oxaloacetate transaminase (GOT) found in higher concentration in liver have been shown to be useful in accessing the presence and degree of hepatocellular cell damage (Pequignot and Jones, 1982). Any damage of liver by introduction of foreign particle, cause elevation of both the enzymes i.e. GOT and GPT (Wroblewski and La Due, 1996).

Serum glutamate oxaloacetate transaminase (SGOT) is an enzyme found mainly in heart muscles, liver cell, skeletal muscles and kidneys. Injury to these tissues results in the release of the enzyme in blood stream. Elevated levels are found in myocardial infraction, cardiac operations, hepatitis, cirrhosis, acute pancreatitis, acute renal diseases, and primary muscle diseases. Decreased levels may be found in pregnancy, Beriberi and Diabetic Ketoacidosis. Serum glutamate pyruvate transaminase (SGPT) is found in a variety of tissues but is mainly found in the liver. Increased levels are found in hepatitis, cirrhosis, obstructive jaundice and other hepatic diseases. Slight elevation of the enzyme is also seen in myocardial infraction (Reitman and Frankel, 1957).

The pathological changes of liver during the treatment period is monitored by some suitable biochemical parameters directly related to liver pathogenesis, e.g. GOT, GPT, alkaline phosphatase, glutamic dehydrogenase, etc. (Boyde, 1968, Brickerstaff and Zimmerman, 1986).

Determination of the activities of serum enzyme in many clinical conditions has assumed an important position in recent years. The activities of many enzymes are now more or less routinely determined for diagnostic purposes. Duration of increased enzyme activity is frequently of diagnostic importance. GOT reaches peak after about 24 to 36 hours (De Ritis *et al,* 1957).

De Ritis *et al.* (1957) gave a ratio on GOT: GPT as a means to distinguish between the predominantly inflammation lesion and necrotic processes. De Ritis ratio less than one indicates inflammation while the ratio greater than one suggested necrosis. The half life of GOT is 50 hours and GPT is 75 hours (Wolf et al, 1973). The normal range of GOT and GPT activity is- GPT: 5-45 IU/L and GOT: 0-42 IU/L (Kaplan, 1993).

The Congo red binding- and DNase agar positive bacterial isolates are found to have an effect in decreasing the body weight of mouse when administered orally (Table 8).

*Enzymatic Titre of the two Enzymes GOT and GPT of test mice fed with bacterial isolates from partially cured fish* - The

GOT and GPT enzyme activity of the test mice fed with *Staphylococcus aureus* and *Streptococcus faecalis* were recorded (Table 9).

**Table 8.** Effect of administration of bacterial isolates to experimental mice on their body weight.

| S.No. | Bacterial Species fed to Experimental mice | Initial Weight (in gram) | Final Weight (in gram) |
|---|---|---|---|
| | **From Partially Cured Fish** | | |
| 1 | Control | 21.32 | 22.51 |
| 2 | *Staphylococcus aureus* | 27.05 | 23.70 |
| 3 | *Streptococcus faecalis* | 21.33 | 20.21 |
| | **From Fermented Fish** | | |
| 4 | *Staphylococcus* sp. | 23.14 | 22.62 |
| 5 | *Streptococcus* sp. | 19.64 | 18.57 |
| 6 | *Bacillus subtilis* | 29.74 | 25.90 |

**Table 9.** GOT and GPT level of test mice fed with *Staphylococcus aureus* and *Streptococcus faecalis.*

| S No. | Bacterial Isolates fed to Experimental Mice | GOT (IU/L) | GPT (IU/L) |
|---|---|---|---|
| 1 | Control | 49.90 | 36.5 |
| 2 | *Staphylococcus aureus* | 90.00 | 105.5 |
| 3 | *Streptococcus faecalis* | 96.00 | 111.0 |

The GOT and GPT level in the experimental mouse fed with *Staphylococcus aureus* and *Streptococcus faecalis* are found to be more than that the normal ratio.

*Enzymatic Titre of the two Enzymes GOT and GPT of test mice fed with bacterial isolates from fermented fish* - The GOT and GPT enzyme activity of the test mice fed with *Staphylococcus* sp., *Streptococcus* sp. and *Bacillus subtilis* were recorded (Table 10).

**Table 10.** GOT and GPT level of test mice fed with *Staphylococcus* sp., *Streptococcus* sp. and *Bacillus subtilis.*

| S No. | Bacterial Isolates fed to Experimental Mice | GOT (IU/L) | GPT (IU/L) |
|---|---|---|---|
| 1 | Control | 33.33 | 21 |
| 2 | *Staphylococcus* sp. | 123.30 | 63 |
| 3 | *Streptococcus* sp. | 95.00 | 66 |
| 3 | *Bacillus subtilis* | 73.34 | 87 |

The GOT and GPT level in all the three bacterial species as recorded in the above table shows higher ration than the prescribed normal one.

The study therefore revealed that the GOT and GPT level are higher than the normal range (GPT: 5-45 IU/L; GOT: 0-42 IU/L) in case of all the five strains. This indicated a possible hepatotoxic nature of these organisms. The administration of these bacterial isolates with the feed also decreases the body weight of the mice compared to the control mouse. Sherlock and Dooley (1993) reported the toxigenic effect of liver cell by various viral, bacterial and fungal strains and monitored by suitable biochemical parameters. However, the presence of *Bacillus subtilis*, *Staphylococcus* sp and *Streptococcus* sp. in the partially cured and fermented fish was due to contamination during processing. Beumer (2001) reported that a small number of *Bacillus* sp. counts in foods are not considered significant. *Staphylococcus aureus* is regarded as a poor competitor and its growth in fermented foods is generally associated with a failure of the normal microflora (Nychas and Arkoudelos, 1990). Pathogenic contaminating organisms like *Bacillus cereus*, *Staphylococcus aureus* and enterobacteriaceae were also detected by Thapa et al. (2004) from Nagri, Hentak and Tungtap. Though there has been no reported case of toxicity or illness due to consumption of fermented fish products in North-East India, gross contamination of fermented fish products in the region suggested that the products need to be investigated clinically (Thapa et al., 2004). Microbial safety hazards in fermented fish are represented by *Staphylococcus aureus* and *Clostridium botulinum*. The presence of *Staphylococcus aureus* and *Clostridium* sp., in traditionally fermented fish seem to be attributed to the contamination of raw materials used during processing. Both microorganisms could be associated with salt used in the processing. However this kind of contamination can be prevented during processing if the processing is carried out in hygienic conditions.

## References


Adams, M. R. and Moss, M. O. (2003). Fermented and Microbial Foods, In: Food Microbiology, 2nd edn. Panima Publishing Corporation, New Delhi. pp 347-348.

Adams, M. R. and Nicolaides, L. (1997). Review of the sensitivity different food borne pathogens to fermentation. *Food Control.* 8: 227-239.

Alur, M. D., Lewis, N. F. and Kumta, U. S. (1971). Spoilage potential of predominant organisms and radiation survivors in fishery products. *Indian J. Exp. Biol.* 9: 48-52.

Banwart, G. J. (1987). Useful Microorganisms, In: Basic Food Microbiology. CBS Publishers & Distributors, New Delhi. pp 453, 482.

Banwart, G. J. (1989). Food spoilage, In: Basic Food Microbiology. CBS Publishers & Distributors, New Delhi. pp 165-67, 442.

Berkhoff, H. A. and Vinal, A. C. (1986). Congo red medium to distinguish between invasive and non invasive *Escherichia coli* pathogenic for poultry. *Avian Dis.* 30: 117-121.

Beumer, R. R. (2001). Microbiological hazards and their control. In: Fermentation and Food Safety, ed. M.R. Adams and M.J.R. Nout, Gaithersburg, Md., USA, Aspen Publishers, Inc. ISBN 0-8342-1843-7. pp 141-157.

Bonnell, A. D. (1994). Quality Assurance in Seafood Processing: A practical Guide, Chapman & Hall, New York.

Boyde, T. R. C. (1968). Glutamic oxaloacetic transaminase isoenzymes. *Biochem. J.* 82 (3): 51.

Brickerstaff, J. R. and Zimmerman, H. J. (1986). Enzyme levels in the diagnosis of hepatic disease. *Amer. J. Gastroenterology.* 40 (2): 387.

Chichcester, C.D. and Graham, H.D. (1973). Microbial safety of fishery products. Academic press, New York, London. pp 118-134.

De Ritis, F., Coltorti, M. and Giusti, G. (1957). Diagnostic value and pathogenic significance of transaminase activity changes in viral hepatitis. *Minerva Med.* 47 (3): 167.

Doyle, M. P. and Decker, M. (1989). Food-borne Bacterial Pathogens. Marcel Decker, New York. pp 36-48.

Fransen, N. G. (1983). Silage fermentation of fish with lactic acid bacteria. *J. Sci. Food-Agris.* 34 (10): 1057-1067.

Frazier, W. C. and Westhoff, D. C. (2005). Miscellaneous Foods. In: Food Microbiology, $4^{th}$ edn. Tata McGraw-Hill Publishing Company Ltd. New Delhi. p. 311.

Garthwaite, G. A. (1997). Chilling and freezing of fish. In: Fish Processing Technology, ed. G.M. Hall, VCH Publisher, New York. pp 93-119.

Hog, M.E., Islam, M.N. and Chakraborty, S.C. (1983). Bacteriological evaluation of hilsa (*Hilsa ilisha*) products in Bangladesh . *Bang. J. Microbiology.* 128-137.

Inglish, V., Richard, R. H., and Wodward, K. N. (1993). Public health aspect of bacterial infections of fish. In: Bacterial Diseases of Fish, ed. Valery Inglish, Ronald Roberts, and Niall Bromage, Halsted Press, New York. pp 284-303.

Kaplan, M. M. (1993). Lab. Tests. In: Diseases of the liver, $7^{th}$ edn., ed. L. Schiff and E.R. Schiff. Philadelphia, Lipincott. pp 108-144.

Knochel, S. and Huss, H. H. (1984). Ripening and spoilage of sugar-salted herring with and without nitrate. I. Microbiological and related chemical changes. *J. Food Technol.*, 19 (2): 203-213.

Kreig, N. R. and Holt, J. G. (1984). Bergy's Manual of Systematic Bacteriology. Vol.1. Baltimore, Williams and Wilkins Co.

McKercher, P. D., Hess, W. R. and Hamdy, F. (1978). Residual viruses in pork products. *Appl. Environ. Microbiol.* 35: 142-145.

Nerquaye-Tetteh, G., Eyeson, K. K. and Tette-Marmon. (1978). Studies on "Bomone"- A Ghanaian fermented fish product. Accra, Ghana, Food Research Institute (CSIR).

Nickerson, J. T. and Sinskey, A. J. (1972). Microbiology of flesh-type foods and eggs. In: Microbiology of Food and Food Processing. American Elsevier Publishing Company, Amsterdam, New York, London. pp 152-158.

Nychas, G.J.E. and Arkoudelos, J.S. (1990). Staphylococci: their role in fermented sausages. *J. Appl. Bacteriol. Symposium Supplement.* 19: 167S-188S.

Okonkwo, P. O. and Nwokolo, C. (1977). Aflatoxin B: Procedures to reduce levels in Tropical Foods. Nutrition Reports International. 17 (3): 587-595.

Pequignot, F. and Jones, J.S. (1982). The clinical significance of transaminase activities of serum. *Amer. J. Med.* 27 (3): 911.

Phithakpol, B. (1993). Fish fermentation technology in Thailand. In: Fish Fermemtation Technology, ed. Lee, C.H., Steinkraus, K.H. and Alan Reily, P.J. pp 155-166. Tokyo: United Nations University Press, ISBN 89-7053-93480-8.

Reilly, P. J. A., Parry, R. W. H. and Barile, L. E. (1990). Post-harvest Technology, Preservation and Quality of Fish in South East Asia. International Foundation for Science, Stockholm. pp. 176-177.

Roberts D, 1982. Bacteria of Public health significance. *in Fish Microbiology*, edited by M H Brown. Applied Science Publishers Ltd., London and New York. Pp 319-386.

Roberts, T. A. and Skinner, F. A. (1983). Food Microbiology Advances and Prospects. Published for the society for Applied Bacteriology by Academic Press. pp 221-243.

Sanni, A. I., Asiedu, M. and Ayernor, G. S. (2002). Microflora and Chemical Composition of Momoni, a Ghanaian Fermented Fish Condiment. *Jr. Food Compos. Anal.* 15 (5): 577-583.

Sherlock, E.M. and Dooley, E.L. (1993). The transaminase tests in liver disease. *Medicine.* 46(4): 197.

Spencer, R. and Hughes (1963). Recent advances in fish processing technology. *Food Manufac.* 38: 407-412.

Tansaupawat S & Komangata M, 1995. Traditional fermented Food of Thailand. *World J. Microbiol.*, 11(2), 253-255.

Tarr, H.L.A. (1954). Microbiological Deterioration of Fish Post Mortem, its Detection and Control. *Bacteriol.Rev.* 18 (3): 1-15.

Thapa N, Pal J & Tamang J P, 2004. Microbial diversity in ngari, hentak, and tungtap, fermented fish products of North-East India. *World J. Microbiol. Biotechnol.*, 20(6), 599-607.

Townsend, J. F. (1971). Fungal flora of South Vietnamese Fish and Rice. *J. Trop. Med. Hyg.* 74 (4): 98100.

Van ver, A. (1965). Fermented and Dry Sea Food Products in South East Asia. In: Fish as Food, Vol.III. ed. Borgstrom, G., Academic Press Inc., New York. pp 237-245.

William, C. Frazier. And Dennis, C. Westhoff. (1995).Contamination, Preservation and Spoilage of Fish. In: Food Microbiology, 4th edn.

Wroblewski, F. and La Due, J. S. (1996). Serum glutamic pyruvic transaminase in hepatic disease, a preliminary report. *Amer. Intern. Med.* 45 (2): 801.

Wu, Y.C., Kimuru, B. and Fujii, T. (2000). Comparision of three culture methods for the identification of *Micrococcus* and *Staphylococcus* in fermented squid shiokara. *Fisheries Sc.* 66: 142-146.

# Chapter 4

# Nutritional Aspects of Fermented Fish

India is blessed with a coast line of 8118 km with an Exclusive Economic Zone (EEZ) of 2.02 million square km and a continental shelf of 0.506 million sq. km. The inland water resources includes 191024 km of rivers and canals; 2.05 million ha of reservoirs; 2.254 million ha of ponds and tanks; 1.3 million ha of oxbow lakes and derelict waters; and 1.24 million ha brackish waters. India has an estimated fish production potential of 8.4 million tons, of which the marine sector forms 3.9 million tons and the inland sector 4.5 million tons. As against this, the current fish production is 5.96 million tons *i.e.,* 2.89 million tons from marine sector and 3.07 million tons from inland sector (Vannuccini, 2003). The per capita fish availability in India is 4.7 kg/year (Laurenti, 2002).

In fact, fisheries sector forms the bread and butter and nutritional source for millions of Indians. Post harvest loss of resources is an area of major concern in this sector and one of the solutions for that is production of value added products; which also can contribute to the improvement of the economic status of the poor. Fisheries sector can play a vital role, as a potential source, in attaining nutritional security in India. The current production of fish in India forms only 71.0% of the total potential and hence there is an ample scope of improvement and thus can be added up to the nutritional security. Demand for fish and fish related products are increasing day by day in our country and reduction in post harvest losses (through curing and other preservation techniques) can make a major contribution to satisfying this demand, improving quality and quantity for consumers and increasing income for the producers. Thus, improvement in the post harvest utilization of fish catches can ensure further nutritional security among a wide range of people in India. Viewing fish primarily as a source of protein has been the dominant perspective in studies on nutritional security. This view is primarily not in the right direction, since it fails to highlight the critical role of fish as a wholesome source of Poly Unsaturated Fatty Acids (PUFA's); minerals such as calcium, phosphorus, iron; vitamins like A, $B_1$, $B_2$, $B_{12}$, D etc.; and trace elements such as iodine and zinc. These attributes makes fish a vital contributor to nutritional security of the most deprived and vulnerable populations.

In the period before World War I, when traditional fermented foods started to attract scientific interest, nothing was known about the role of vitamins, antibiotics, or amino acid requirements in animals and man. These requirements have become a new and importance area of research in traditional foods made by fermentation.

The general aims of food technology are to exploit natural food resources as efficiently and profitably as possible. Adequate and economically sound processing, prolongation of shelf life by preservation and optimization of storage and handling, improvement of safety and nutritive value, adequate and appropriate packaging, and maximum consumer appeal are key prerequisites to achieving these aims. In technologically developed regions, the crafts of baking, brewing, wine making, and dairying have evolved into the large-scale industrial production of fermented consumer goods, including cheeses, cultured milks, pickles, wines, beers, spirits, fermented meat products, and soy sauces. In contrast, many of the traditional indigenous foods lack this image; fermented foods are often considered as backward or poor man's food. Factors contributing to such lack of appeal include inadequate grading and cleaning of raw materials, crude handling and processing techniques, and insufficient product protection due to lack of packaging knowledge. From nutritionist's point of view, many traditional starchy staples are deficient in energy, protein, and vitamins. Variable sensory characteristics (quality) and lack of durability (shelf life) reduce convenience to consumer; time needs to be spent selecting products require frequent purchasing and result in increased wastage. In addition, ungraded heterogenous products, inconvenient unpacked bulk foods, or unattractive presentation inhibit consumers to develop regular purchasing attitude. The contrast as outlined above serves as a general guideline to the major targets for upgrading the present status of traditional indigenous fermented foods. However, these indigenous fermented foods provide an appropriate basis for development of a local food industry, which not only preserves the agricultural produce but also stimulates and supports agroindustrial development. Moreover, fermented food products have many nutritional advantages, which surpass western-style fast food and processed foods.

Traditional Foods are the treasure box of resources for science and technology. The way they are produced, the local substrates used in the countries where they are developed such as plants and animals and the presence of diverse microorganisms in these foods are good and novel resources for food industry technology development. Microbes and fermented products, typical of these traditional foods, have unique functions for human health, and they are good targets for the research of physiologically functional foods.

Fish is highly perishable food item. Although the need for preservation is not so pressing in the extremely cold Arctic regions or the temperate region, the scenario is quite different in the tropical climate like in India. The hot climate of this region favour spoilage as a result of which the landed fish undergoes rigor mortis within 6-8 hours, after which decay is eminent. In the country like India it is not affordable to waste animal protein by the way of spoilage. Therefore, it greatly generates the requirement of preservation of such perishable items, ensuring availability of fish in the market throughout the year. One such means is the process of fermentation. On the other hand bio-enrichment of food substances by traditional fermentation technique with protein, essential amino acids and vitamins enhances nutritive value of the raw material. This has high significance for developing countries where majority of the people are unable to afford to have commercially available expensive fortified nutritive foods.

Fermented food products form an intrinsic part of the diet of the tribal peoples in Northeastern India. Tungtap which is prepared by Khasi tribe of Meghalaya is most popular food item among this tribe. The nutritive value of this fermented food product has been studied and when compared with its non-fermentative counterparts the former was found to be rich in its quality in terms of its physico-chemical and biochemical profile. The fresh *Puntius sarana* showed a moisture content of 73.63% which is abruptly came down to 12.82% in case of partially cured fish (Baishya et al., 2007) and in the fermented product it was recorded to be 17.53%. The low content of moisture in the partially cured fish and fermented products therefore, ensures preservation. The moisture content of the fresh Cod fish is 78 to 84% while in the Haddock it is 79 to 84%. The Cod fish is cured by salting and in the salted Cod the moisture content reduces to 33%. In case of Haddock, curing is done by smoking and in the smoked Haddock the moisture content reduces to 72%. Moreover, pH of the fermented fish (Tungtap) also found to be acidic and recorded to be 5.1, which was 7.82 in case of fresh fish and 5.9 in partially cured fish (Baishya *et al.,* 2007). This is due to the production of acid or antimetabolites by some group of bacteria during fermentation. The low pH does not allow the growth of many microorganisms. While pH can be used as a simple scientific indicator of sourness of the product, the evaluation for total acidity ca only be used as an indicator to quantify the concentration of hydrogen ions to be dissociated from any titrated acid. Tungtap, however, showed a high ash content of 30.32%, which was nearly same as reported by Murugkar and Subbulakshmi (2006) in case of fermented *Puntius sophore,* was recorded to be 32.2g/100g. Sarojnalini and Viswanath (1994) also reported the ash content of ngari, a fermented fish product of Manipur to be 11%. The XRF analysis for qualitative estimation of mineral content in the samples showed that Zn, Fe, K and Ti are the common minerals recorded in fresh, partially cured and fermented fish samples besides some other micro and macro elements like Ca, S, Mg, Mn, Cu, etc. (Table 11). There are reports that in tungtap, the process of drying and further fermentation weakens the bones considerably, almost dissolving them in the flesh portion. This could be the reason for high calcium levels of 5040mg/ 100g and phosphorus levels 1930mg/100g in the product as reported by Murugkar and Subbulakshmi (2006).

**Table 11.** The data of physicochemical analyses of fresh, partially cured and fermented fish (*Puntius sarana*).

| Sample Type | Moisture Content (%) | Ash Content (%) | Titrable Acidity (%) | pH | Mineral Content by XRF |
|---|---|---|---|---|---|
| Fresh Fish | 73.63 | 20.34 | 1.61 | 7.82 | Zn, Fe, K, Ti, Ca, S, Mg, Mn, Cu & Si |
| Partially Cured Fish | 12.82 | 29.59 | 3.32 | 5.90 | Zr, Zn, Sr, Fe, Ti, K, Ca, Cl, S, P, Si & Mg |
| Fermented Fish | 17.53 | 30.32 | 4.53 | 5.10 | Zn, Fe, K, Ti, Mn & Cu |

The protein contents of the fermented foods were marginally higher than their unfermented counterparts. The reason could be due the presence of microorganisms which grow during the process of fermentation as also reported by Murugkar and Subbulakshmi (2006). The protein content of fermented fish sample was found to be 48% (Table 12). Another study examining the crude protein of the 1 tung tap a variable range from as low as 1% to as high as 14% (Mizutani et al., 1992). The total lipid in the partially cured fish was found to be higher (11.25%) than the fermented fish sample which was recorded to be 5.45%. Sarojnalini and Viswanath (1994) found that *Puntius sophore* used in a traditional fermented dish preparation (ngari) in Manipur contained 45% protein and 19% lipids. Low total carbohydrate content was recorded in the fresh fish (*Puntius sarana*) while it was quite insignificant in case of partially cured and fermented fish. The reason may be that the microorganisms present in the fish, used up the sugar of the host and carried out fermentation for their metabolic processes. There is, however, increase in total free amino acid content (0.05mg/ml) as recorded in the fermented fish sample and thin layer chromatography of the sample extracts for qualitative estimation of amino acids could also detect almost all the essential amino acids in the fresh, partially cured and fermented fishes indicating thereby a good nutritive value of the product. Murugkar and Subbulakshmi (2006) reported low fibre content (0.4g/100g) in tungtap. They also reported that in tungtap, the process of drying and further fermentation weakens the bones considerably, almost dissolving them in the flesh portion. This reason was attributed to their calcium levels of 5040mg/100g and phosphorus levels of 1930mg/100g in the product. One serving (50g fresh weight) of tungtap contained 13.5g of protein, 6.5g of fat, and 1657mg of calcium, taking care of the requirement of 27% of protein, 21% fat, and upto 4 times the requirement of calcium for the day.

**Table 12.** Biochemical analyses of fresh, partially cured and fermented fish, *Puntius sarana*.

| Sample Type | Protein Content (%) | Total Carbohydrate (%) | Reducing Sugar (%) | Total Lipids (%) | Total Free Amino Acid (mg/ml) |
|---|---|---|---|---|---|
| Fresh Fish | 31.67 | 0.03 | 0.11 | 8.56 | 0.02 |
| Partially Cured Fish | 44.56 | - | 0.04 | 11.25 | 0.05 |
| Fermented Fish | 48.00 | - | 0.02 | 5.45 | 0.05 |

'-' implies that no detectable quantity was recorded.

Looking into the nutritive profile of the Tungtap it can be assumed that this type of fermented food product can really fulfill the requirement of the people especially in a developing country like India. There are a number of nutritional diseases in the developing world today. **Kwashiorkor** is one such nutritional disease caused due to deficiency of protein; **Marasmus**, caused by a combination of protein and calorie deficiencies are found in large number of children between ages of 1 and 3 years in the developing world. The immune systems of the children are impaired on restricted protein diets. Such children often develop diarrhoea and the mothers place them on restricted diets such as rice broth. Since the children are already deficient in protein, they quickly develop kwashiorkor with tissue edema, bloated abdomens, susceptibility to infection and changes in hair pigments. Other nutritional diseases common in the developing world are xeropthelmia; childhood blindness due to deficiency in Vitamin A; beri-beri due to thiamine deficiency including infantile beri-beri where sudden death may occur from heart failure in children being nursed by mothers deficient in thiamine; pellagra due to niacin deficiency, riboflavin deficiency; rickets caused by Vitamin D deficiency; anemia due to Vitamin B-12 deficiency or insufficient iron in the diet (Jelliffe, 1968).

## References

Baishya, D. Pandy, M. and Deka, M. (2007). Certain physicochemical and biochemical parameters of *Puntius sarana* (Ham.) and its antibacterial activity. *Asian J. Microbiol. Biotechnol. EnvSc.* 9(4): 845-850.

Laurenti, G. (2002). Fish and Fishery Products: World apparent consumption statistics based on food balance sheets. FAO Fisheries Circular No. 821.

Mizutani, T., Kimizuka, A., Riddle, K. and Isige, N. (1992). Chemical components of fermented fish products. *J. Food Comp. Anal.* 5(2): 152-159.

Muruhgkar, D. Agrahar, Subbulakshmi, G. (2006). Preparation Technique and Nutritive Value of Fermented Foods from Khasi tribes of Meghalaya. *Ecology of Food and Nutrition.* 45: 27-38.

Sarojnalini, C. and Viswanath, W. (1994). Composition and nutritive value of sun-dried *Puntius sophore. J. Food Sci. Technol.* 31: 480-483.

Vannuccini, S. (2003). Overview of Fish Production, utilization, consumption and trade. Fishery Information, Data and Statistics Unit, Food and Agriculture Organization, Rome, Italy.

# Chapter 5

# Probiotics and Fermented Fish

In 1908, Metchnikoff, the father of modern Immunology, wrote "Prolongation of Life" and created the incredible ongoing revolution known as the "Probiotic Revolution". He proposed that "Lactic Acid Bacteria" (LAB) could render a great service in the fight against intestinal putrefaction and might postpone and ameliorate old age. Till then today, some 95 years later, dozens of the worlds most brilliant minds carry on the astonishing work that has scientifically been proven to generate restorative and healing effects in the body.

In the late $19^{th}$ century, microbiologists identified microflora in the gastrointestinal (GI) tracts of healthy individuals that differed from those found in diseased individuals. These beneficial microflora found in the GI tract were termed probiotics. The term 'Probiotics' is originally a Greek word meaning 'for life'. Parker (1974) for the first time used this word. He defined Probiotics as "organisms and substances, which contribute to intestinal microbial balance". But the most accepted definition of Probiotics was given by Fuller (1989). He redefined Probiotics as "a live microbial feed supplement, which beneficially affects the host animal by improving its microbial balance". Fuller's definition since been broaden to state "a probiotic is a mono or mixed culture of live microorganisms which, when applied to animal or man, affect the host beneficially by improving the properties of the indigenous microflora". This definition stresses the importance of live microorganisms that improve the health status of either man or animal and that occur in the mouth, gastrointestinal tracts (GIT), or upper respiratory or urogenital tracts (Havenaar et al, 1992).

To understand how probiotics work, it is important to understand a little about the physiology, microbiology of GI tract and the digestive process. The digestive process begins as soon as food enters the mouth and to stomach, the microbes present in the GI tract have the potential to act in a favourable, a deleterious or a natural manner. Microbes in small intestine and in large intestine complete the digestion process.

Certain intestinal microbes are known to produce vitamins and they are nonpathogenic, their metabolism is non-putrefactive, and their presence is correlated with a healthy intestinal flora. The metabolic end products of their growth are organic acids (lactic and acetic acids) that tend to lower the pH of the intestinal contents, creating conditions less desirable for harmful bacteria. Probiotics may also influence other protective functions of the intestinal mucosa including synthesis and secretion of antibacterial peptides, mucins. The GI tract also serves as a large mucosal surface that bridges the gap between 'inside the body' and 'outside the body'. Along this mucosal surface, microbes and foreign antigens colonizing or passing through the GI tract interact with important components of immune system. This interaction serves to prime or stimulate the immune system for optimal functioning. Normal microbial inhabitants of the GI tract also reinforce the barrier function of the intestinal lining, decreasing 'translocation' or passage of bacteria or antigens from the intestine into the blood stream. The function has been suggested to decrease infections and possibly allergic reactions to food antigens (Parvez *et al.,* 2006).

Ideally, a microorganism should meet a number of predefined criteria in order to be considered as probiotic. Probiotic microorganisms must to be proven to be safe and efficacious in humans, following specific protocols for their isolation. Adherent probiotic strains are desirable because they have a great chance of becoming established in the GIT, thus enhancing their effect (Lee, *et al,* 1999). All probiotic strains should have generally recognized as safe (GRAS) status, be nonpathogenic, and cause no adverse effects to the recipient (Prajapati and Nair, 2003). The selection criteria for a lactic acid bacteria to be used as probiotic include the ability to: (i) exert a beneficial effect on the host; (ii) withstand into a foodstuff at high cell counts, and remain viable throughout the shelf-life of the product; (iii) withstand transit through GIT; (iv) adhere to the intestinal epithelium cell lining and colonize the lumen of the tract; (v) produce antimicrobial substances towards pathogens; and (vi) stabilize the intestinal microflora and be associated with health benefits (Parvez *et al,* 2006). Bacterial lactase hydrolyses lactose that aids in lactose digestion. Probiotics resist the colonization of enteric pathogens; alter the intestinal conditions to be less favourable for pathogenicity (pH, short chain fatty acids, bacteriocins), adhere to intestinal mucosa, interfering with pathogen attachment to the intestinal epithelial cells.

Reports are there on alteration of toxin binding sites, mutagen binding sites, carcinogen deactivation, inhibitions of carcinogen producing enzymes of probiotic organism. Products like lactic acid can inhibit the growth of *Helicobacter pylori* (Sander and Huis, 1999).

Since the bacteria in general are sensitive to low pH values in stomach (Conway *et al.*1987, Berrada *et al.* 1990), it is most important that probiotics are consumed within a food matrix. Bacteria can survive acidic conditions *in vitro* when inoculated onto the surface of solid food, whereas the same level of acidity is lethal to the inoculum in an acidified broth environment. Milk has been shown to be an excellent vehicle for probiotic bacteria probably due to its high buffering capacity (Saxelin, 1996). Fish flesh acts also as a buffer in an acidic environment and therefore may also protect bacteria from hostile environment. The protective effect of some solid foods may be due to the raising the pH of microenvironment of the bacteria on the surface of the food. Ehrmann et al. (2002) suggested that for being a probiotic agent the organism should be able to tolerate low pH and bile acids. Certain other line of studies suggest to assess the effect of temperature, agitation (oxygen requirement), and certain inhibitory substances like phenol and salt (NaCl) on the growth of the organism in order to characterize them for probiotic parameters (Reque et al., 2000).

*The Lactic Acid Bacteria and its role in fermentation:* The main sources of the probiotic products are bacteria. The bacteria found in this products often shows certain positive health effects such as quicker recovery from certain types of diarrhea, enhanced immune function, decrease in probability of occurring cancer and possibly lower of blood cholesterol level (Suverna and Boby, 2005). One of the significant group of probiotic organisms are Lactic Acid

Bacteria (LAB).They are of common occurrence in fermented food products.

Lactic acid bacteria are those that produce lactic acid as the sole product or the major acid from the energy yielding fermentation sugars. Traditionally it has been easy to define the group but in recent years, it has become more difficult so that opinions differ as to what should be called a lactic acid bacterium. They can be broadly defined as gram positive; anaerobic, micro-aerophilic or aero-tolerant bacteria, either rod or coccus shaped catalase negative, and fastidious in their growth requirements; many need B vitamins to grow. This definition can include an endospore forming bacterium (Sporolactobacillus), and a motile species but none of these are important for food production and preservation (Lee, 2003).

The major genera of LAB of importance in the food industry for food preservation and production (some cause food spoilage) are Bifidobacterium, Enterococcus, Lactobacillus, Lactococcus, Leuconostoc, Oenococcus, Pediococcus, Streptococcus and Tetragenococcus. They are important in the preservation and production of many different foods for humans as well as for animals. These foods include milk and dairy products, meat products, vegetable products, and forage crops for animals. When these lactic acid bacteria ferment sugars, they produce varying amounts of lactic acid. This is due to two different metabolic pathways by which they may metabolize the sugars. The pathway results in different metabolic bi-products. One pathway results in homolactic fermentation, when the other pathway leads to heterolactic fermentation. Consequently, the lactic acid bacteria are generally categorized on the basis of these fermentation patterns (Lee, 2003). The genera, Streptococcus and Pediococcus are homofermentative, the Leuconostocs are heterofermentative and Lactobacilli include both homo- and heterofermentative types. The homofermenters convert carbohydrates primarily to lactic acid, while the heterofermenters produce lactic acid and substances such as acetic acid, ethyl alcohol and carbon dioxide. This means that the homofermenters use primarily one metabolic pathway, whereas, the heterofermenters use more than one pathway (Banwart, 1987).

It is meaningful for food processing to select strains with outstanding properties by screening for the biochemical properties of lactic acid bacteria from functional qualities of metabolism and cultivation. From this viewpoint, the screening for chloride-resisting lactobacillus and acid-resisting lactobacillus, and moreover, the strains with outstanding functional qualities such as acid productivity, proteolytic effect, fat resolving effect, fragrance productivity, and antimicrobial component productivity have been in progress. Furthermore, in a research into bacteria aimed to increase lactic acid bacteria's range of use, it is essential to screen for lactic acid bacteria with special saccharide fermentativeness or heat-resisting effect in order to search for the ones that are applicable to a wide range of fermented materials. These researches are all vital in the sense that functional properties in lactic acid bacteria improve preservative effect and add flavor and taste. Geographically, Southeast Asia belongs to the tropical region and various traditional fermented foods—salted in most cases—are utilized there due to the climatic peculiarity of the region. It is presumed that those lactic acid bacteria with unique, strong fermentation activity exist in natural fermented foods originating in this unique, tropical environment.

The use of probiotics in fermentation has numerous advantages. The fermentation acts to retain and optimize microbial viability and productivity, while simultaneously preserving the probiotic properties. The fermented fish and other fermented products are believed to have medical value i.e. probiotic and products and the market for such products are developing, with most activity in developed countries, in particular in Europe, Japan and the United States (Stanton, *et al,* 2001). Lindgun and Dobrogosz (1990) reported that LAB can be used to compete with the gastrointestinal microbes found in fishes. Authentic lactic-fermented fish products have to include as an ingredient and exogenous source of fermentable carbohydrate. Considerable variation in recipes has been noted but production is governed by two general principles: the higher the salt content of the product, the longer the production process takes but the better the product's keeping qualities; and the higher the level of added carbohydrate, the faster the fermentation and the more acidic the flavour (Adams and Moss, 2003).

Lactic acid fermentation includes those in which the fermentable sugars are converted to lactic acid by lactic acid organisms. This single category is responsible for processing and preserving vast quantities of human food and insuring its safety. Fish products fermented with lactic acid are generally processed out of fresh water species of fish. Fermentation with lactic acid is quicker than that with salt alone (6-18%). Their processing involves mixing the fish with salt and a carbohydrate such as cooked rice. The keeping quality as well as the extent of acid fermentation depends on the amount of salt and carbohydrate used. Carbohydrate is used as a good source for lactic acid producing bacteria. 'Pla-ra' is a prominent lactic acid fermented product of Thailand. Consistency of Pla-ra varies from dark brown liquid to a partly fried fish pieces and the finished product contain microflora dominated by yeast and LAB (Reilly et al., 1990). Philippine's 'balao balao' is a lactic acid fermented rice/shrimp mixture prepared by mixing boiled rice, raw shrimp and solar salt (about 3%w/w), packing in anaerobic container and allowing the mixture to ferment over several days or weeks (Arroyo et al, 1977; Steinkraus, 1983).

The phenomenon of a lactic acid bacterium inhibiting or killing closely related and food poisoning organisms when in a mixed culture has been observed for more than sixty years. This antagonism by lactic acid bacteria has received extensive study, and several reviews are reported by different workers (Appleton et al, 1981; Hejkal and Gerba, 1981; Smith et al, 1983). The many situations in which LAB dominate can be explained by the production of lactic acid that lowers the pH and prevents the growth of many acid sensitive bacteria. However, in addition many lactic acid bacteria produce substances that kill other bacteria. The most important group is the bacteriocins. Bacteriocins are antimicrobial peptides, proteins or protein complexes that particulary inhibit closely related strains. Many bacteriocins produced by lactic acid bacteria are also active towards food spoilers and/or food-borne pathogenic bacteria such as other lactic acid bacteria, *Bacilli, Clostridia, Staphylococci* and *Listeriae*. In addition, many lactic acid bacterium bacteriocins are active in a wide temperature and pH range. They may hence be applied in future food preservation (to replace chemicals such as nitrite, nitrate, sulfite, etc.), to produce consumer-friendly, natural and safe food. The first bacteriocin described was **nisin,** discovered around 1947, and it is now produced commercially and incorporated in many foods. Hydrogen peroxide is well known for its antimicrobial properties. Since LAB possess a number of flavoprotein oxidases but lack the degradative enzyme catalase, they produce hydrogen peroxide in the presence of oxygen. This will confer some competitive advantage as they have been shown to be less sensitive to its effects than some other bacteria. Accumulation of hydrogen peroxide has been demonstrated in some fermented foods but its effects are, in general, likely to be slight. Lactic acid fermentations are essentially anaerobic processes so hydrogen peroxide formation will be limited by the amount of oxygen dissolved in the substrate at the start of fermentation. It may be, however, that at this critical initial stage of a fermentation hydrogen peroxide production provides an important additional selective advantage (Lee, 2003).

Different species of the lactic acid bacteria that are mostly used as probiotic cultures has been studied in molecular detail in order to identify the isolated specified target of the functional strains. Many different genotyping techniques may be applied to LAB as tools for either species identification or differentiation of strains to the clonal level. The major advantages of these DNA-based typing methods lie in their discriminatory power (Farber, 1996) and in their universal applicability. Closely related strains with similar phenotypic features may now reliably be distinguished by DNA-based techniques. Molecular typing methods applicable to probiotic LAB include plasmid profiling, restriction enzyme analysis (REA), pulse-field gel electrophoresis (PFGE), randomly amplified polymorphic DNA (RAPD), and ribotyping. The use of PCR assay (polymerase chain reaction) method with the 16S rRNA analysis has been studied in many works on identification of organisms isolated from chicken host, human faeces and animal feed (Likotrafiti, et al, 2004; Lofstrom, *et al,* 2004; Zhu, *et al,* 2002; Lu, *et al,* 2003).

The bacterial species belonging to the genus Lactobacillus from partially cured and fermented *Puntius sarana* were found to tolerate bile salt upto 6%; NaCl upto

4% but grew poorly at 8%; an acidic pH upto 1 and 2, which can be considered as good probiotic properties. Out of twelve isolates, two Lactobacillus species viz. *Lactobacillus lactis* (BVS14) and *Lactobacillus fermentum* (BI26) isolated from skin and intestine of the partially cured fish respectively have on the other hand shown luxuriant growth compared to other members (Table 13, 14, and 15). Bile tolerance has been described as an important factor for the survival and growth of LAB in the intestinal tract (Gilliland and Walker, 1990). Moreover, tolerance to acidic condition is another important characteristics of probiotic bacteria in order to reach the intestine the organism has to pass through the stomach where the pH level is as low as 1- 2.5. Since the pH of human gastrointestinal tract ranges from 0.9 -1.5 and the two bacterial isolates viz. *L. fermentum* and *L. lactis* could thrive at a low pH level of 1 – 2, so these two isolates were extreme acidophilic in nature.

**Table 13.** Effect of bile salt concentrations on growth of bacterial isolates from *Puntius sarana*.

| Isolate No. | Bacterial Species | Absorbance at different bile salt concentrations (λ = 540nm) | | |
|---|---|---|---|---|
| | | 4% | 5% | 6% |
| BDS1 | *Lactobacillus fermentum* | 0.103 | 0.065 | 0.051 |
| BDS2 | *Lactobacillus ruminis* | 0.226 | 0.131 | 0.121 |
| BDS8 | *Lactobacillus casei* | 0.285 | 0.144 | 0.117 |
| BVS14 | *Lactobacillus lactis* | 0.266 | 0.380 | 0.202 |
| BVS35 | *Lactobacillus casei* | 0.232 | 0.226 | 0.119 |
| BVS 40 | *Lactobacillus fermentum* | 0.183 | 0.018 | 0.074 |
| BG20 | *Lactobacillus fermentum* | 0.219 | 0.121 | 0.108 |
| BI26 | *Lactobacillus fermentum* | 0.572 | 0.428 | 0.221 |
| FFG1 | *Lactobacillus* sp. | 0.117 | 0.107 | 0.016 |
| FFS4 | *Lactobacillus* sp. | 0.018 | 0.012 | 0.050 |
| FFS17 | *Lactobacillus coryneformis* | 0.129 | 0.117 | 0.022 |
| FFS18 | *Lactobacillus Plantarum* | 0.156 | 0.110 | 0.020 |

**Table 14.** Effect of acid concentrations on growth of bacterial isolates from *Puntius sarana.*

| Isolate No. | Bacterial Species | Absorbance at different acid concentrations (λ = 540nm) | | | |
|---|---|---|---|---|---|
| | | pH1 | pH2 | pH3 | pH4 |
| BDS1 | *Lactobacillus fermentum* | 0.001 | 0.002 | 0.003 | 0.042 |
| BDS2 | *Lactobacillus ruminis* | 0.003 | 0.007 | 0.017 | 0.249 |
| BDS8 | *Lactobacillus casei* | 0.010 | 0.020 | 0.050 | 0.395 |
| BVS14 | *Lactobacillus lactis* | 0.110 | 0.120 | 0.127 | 0.116 |
| BVS35 | *Lactobacillus casei* | 0.001 | 0.001 | 0.010 | 0.093 |
| BVS 40 | *Lactobacillus fermentum* | 0.003 | 0.005 | 0.006 | 0.303 |
| BG20 | *Lactobacillus fermentum* | 0.020 | 0.070 | 0.074 | 0.545 |
| BI26 | *Lactobacillus fermentum* | 0.050 | 0.053 | 0.008 | 0.220 |
| FFG1 | *Lactobacillus* sp. | 0.030 | 0.030 | 0.022 | 0.182 |
| FFS4 | *Lactobacillus* sp. | 0.002 | 0.003 | 0.026 | 0.258 |
| FFS17 | *Lactobacillus coryneformis* | 0.002 | 0.040 | 0.048 | 0.221 |
| FFS18 | *Lactobacillus Plantarum* | 0.004 | 0.040 | 0.051 | 0.218 |

**Table 15.** Effect of salt (NaCl) concentrations on growth of bacterial isolates from *Puntius sarana.*

| Isolate No. | Bacterial Species | Absorbance at different NaCl concentrations (λ = 540nm) | | |
|---|---|---|---|---|
| | | 4% | 6% | 8% |
| BDS1 | *Lactobacillus fermentum* | 1.220 | 0.922 | 0.607 |
| BDS2 | *Lactobacillus ruminis* | 1.180 | 0.801 | 0.666 |
| BDS8 | *Lactobacillus casei* | 1.110 | 0.821 | 0.625 |
| BVS14 | *Lactobacillus lactis* | 2.210 | 0.823 | 0.592 |
| BVS35 | *Lactobacillus casei* | 2.400 | 0.706 | 0.507 |
| BVS 40 | *Lactobacillus fermentum* | 1.420 | 0.722 | 0.455 |
| BG20 | *Lactobacillus fermentum* | 1.060 | 0.832 | 0.520 |
| BI26 | *Lactobacillus fermentum* | 2.160 | 0.776 | 0.607 |
| FFG1 | *Lactobacillus* sp. | 1.020 | 0.711 | 0.433 |
| FFS4 | *Lactobacillus* sp. | 1.340 | 0.782 | 0.518 |
| FFS17 | *Lactobacillus coryneformis* | 1.620 | 0.912 | 0.421 |
| FFS18 | *Lactobacillus Plantarum* | 1.510 | 0.906 | 0.421 |

Inhibition of the growth of microorganisms is again considered as a probiotic property. Therefore, all the twelve Lactobacillus members were tested against two pathogenic bacterial isolates viz. *E. coli* and *Salmonella* sp. In order to examine their anti- bacterial activity. It was observed that all the twelve bacterial isolates have shown zone of inhibition against *E. coli* and *Salmonella* sp. Thereby supporting its probiotic nature (Table 16, Fig 11 and 12). There are reports describing that the inhibition of microbial growth resulted from the presence of lactic acid produced, or due to the production of other antimicrobial compounds showing inhibitory properties (Hose and Sozzi, 1991). LAB are often inhibitory to other microorganisms and is the basis of their ability to improve the keeping quality and safety of many food products.

**Table 16.** Zone of inhibition of growth shown by Lactobacillus isolates against *E. coli* and *Salmonella* sp.

| Isolate No. | Bacterial Species | Diameter of Zone of Inhibition (in mm) | |
|---|---|---|---|
| | | ***E. coli*** | ***Salmonella* sp.** |
| | **Control** | - | - |
| BDS1 | *Lactobacillus fermentum* | 1.9 | 1.3 |
| BDS2 | *Lactobacillus ruminis* | 1.8 | 1.3 |
| BDS8 | *Lactobacillus casei* | 1.8 | 1.2 |
| BVS14 | *Lactobacillus lactis* | 1.9 | 1.3 |
| BVS35 | *Lactobacillus casei* | 1.7 | 1.0 |
| BVS 40 | *Lactobacillus fermentum* | 1.8 | 1.2 |
| BG20 | *Lactobacillus fermentum* | 1.8 | 1.0 |
| BI26 | *Lactobacillus fermentum* | 1.9 | 1.3 |
| FFG1 | *Lactobacillus* sp. | 1.7 | 1.0 |
| FFS4 | *Lactobacillus* sp. | 1.7 | 1.0 |
| FFS17 | *Lactobacillus coryneformis* | 1.9 | 1.3 |
| FFS18 | *Lactobacillus Plantarum* | 1.8 | 1.2 |

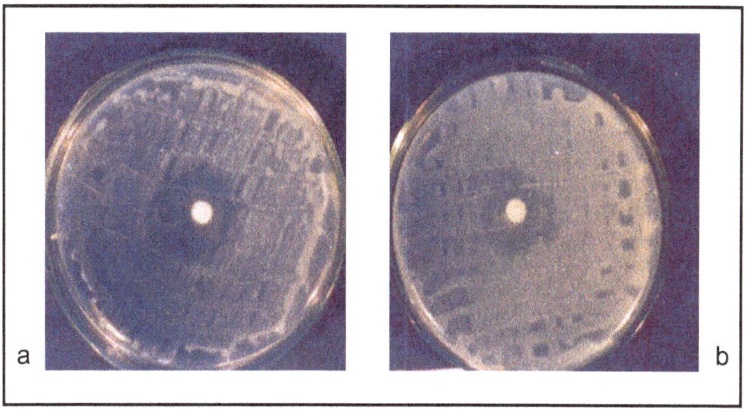


**Fig 11**: Zone of growth inhibition shown by *Lactobacillus fermentum* (BI26) against a) *E. coli* and b) *Salmonella* sp.

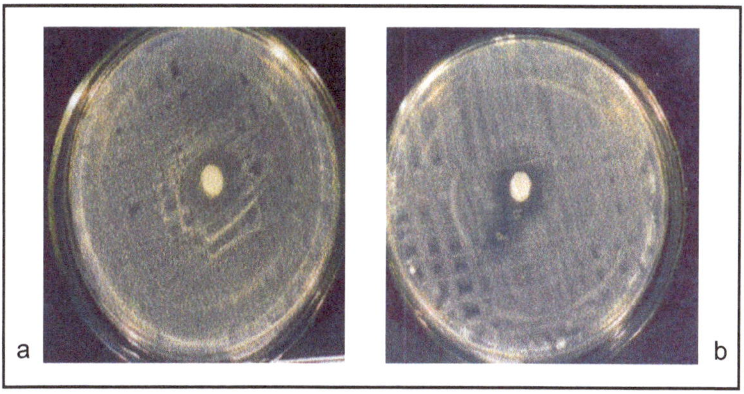


**Fig 12**: Zone of growth inhibition shown by *Lactobacillus lactis* (BVS14) against a) *E. coli* and b) *Salmonella* sp.

In recent years, there has been a considerable increase in studies of the natural antimicrobial compounds on and in food produced by lactic acid bacteria (LAB), referred as bioprotective cultures. Bioprotective cultures may act as starter cultures in the food fermentation process or they may protect foods without any detrimental organoleptic changes.

In a number of fish products, particularly lightly preserved fish products, bacteriological problems (e.g. *L. monocytogenes*) cannot be solved as the bacteria in question will be present and will grow under the preservation conditions used. Therefore, alternative, consumer-friendly preservation techniques must be developed. Strategies used for other foods have included the use of protective cultures, typically lactic acid bacteria, or use of naturally occurring antibacterial compounds like enzymes. In case of fermented fish products also investigations are on in search of enzymes and peptides to be used as alternatives.

Whilst biopreservation and fermentation of seafood typically rely on the antibacterial effect of lactic acid bacteria, the interaction (and antibacterial effect) of certain Gram-negative bacteria is also studied. This is, in fresh fish, related to the Spoilage area, but the antibacterial properties of certain Gram-negative bacteria may be used as disease prevention (probiotics).

*Survivability of bacterial isolates in the intestinal environment:* To be considered as probiotic organism it should adhere to the intestinal epithelium cell lining and colonize the lumen of the tract. In a study to examine the probiotic nature of the bacterial isolates that were isolated from partially cured and fermented *Puntius sarana*, they were study *in vivo* in order to observe their efficacy to thrive in the intestinal environment of mice. The experiment was performed on adult albino mice of the same age and same sex (male).

The bacterial strains were grown in MRS broth at 37 $^{0}$C for 48-72 hours. The bacterial cells were harvested by centrifuging at 10000 rpm for 10 minutes. After centrifugation the bacterial pellets was taken out aseptically. These pellets were then washed with sterile normal saline followed by centrifugation. The pellets thus obtained were mixed with normal feed of the mouse as follows. On the first day 0.1g of the pellets were mixed with 4g of normal feed, second day 0.2g of the pellets were mixed with 7g of the normal feed, third day 0.3g of pellets with 8g of the feed, fourth day 0.4g of pellets with 9g of feed and on fifth day 0.5g of pellets were mixed with 10g of normal feed. The different dosage of feeds thus prepared was fed to the respective mice. One mouse was taken as control and fed with only normal feed. The bacterial strains were administered to the experimental mice one time per day for continuous five days. During this period the mice were monitored carefully for certain phenotypic characters like body colour, eye colour, behavioural change and increase or decrease in their body weight. Faecal matter was also compared to that of the normal one and loss or gain in the body weight was recorded. After fifth day, the next two days the mice were fed with only sterilized normal diet without addition of bacteria so as to avoid entry of any external microbes. After $7^{th}$ day, the mice were kept starved for next two days to get rid of the faecal content of intestine. On the $10^{th}$ day they were sacrificed along with the control one and the intestine was taken out by dissecting under aseptic conditions.

The intestines of the individual mice were then inoculated into MRS broth and incubated at 37 $^0$C for 48 hours. After incubation, pure colonies were isolated by pour plate technique and sector streaking method under anaerobiosis. This was followed by colony characteristics, microscopic examination and various biochemical tests as. The re-isolated colonies were then identified consulting the Bergey's Manual of Determinative Bacteriology (Buchanan *et al.*, 1975).

During the study it was observed that there was no change in the eye colour, faecal matter or any behavioural change among the experimental mice. Moreover, there was an increase in the body weight of the mice fed with Lactobacillus members.

This suggests the beneficial effect of these bacterial members inside the host system. Further, all the bacterial members could be recovered from the intestine of the individual mice. This is further suggesting the probiotic nature of these bacterial isolates. Shah (2001) reported that adherence is one of the most important selection criteria for probiotic bacteria. Thapa *et al.* (2004) also reported strains of LAB isolated from ngari, hentak and tungtap showed high degrees of hydrophobicity indicating that the potential adhesion to gut epithelial cells of human intestine advocating their 'probiotic' character.

## References

Adams, M. R. and Moss, M. O. (2003). Fermented and Microbial Foods, In: Food Microbiology, 2nd edn. Panima Publishing Corporation, New Delhi. pp 347-348.

Appleton, H., Palmer, S. R. and Gilbert, R. J. (1981). Food borne gastroenteritis of unknown etiology: A virus infection. *Brit. Med. J.* 282: 1801-1802.

Arroyo, P. T., Ludovico-Pelago, L. A., Solidum, H. T., Chiu, Y. N., Lero, M. and Alcantara, E. E. (1977). Studies on rice-shrimp fermentation: balao balao. *Phil. J. Food Sci. and Tech.* 2: 106-125.

Banwart, G. J. (1987). Useful Microorganisms, In: Basic Food Microbiology. CBS Publishers & Distributors, New Delhi. pp 453, 482.

Buchanan, R. E. and Gibbons, N. E. (1975). Bergey's Manual of Determinative Bacteriology, 8th edn. The William and Wilkins Company, Baltimore.

Ehrmann, M. A., Kurzak, P., Bauer, J. and Vogel, R. F. (2002). Characterization of Lactobacillus towards their use as probiotic adjunct in poultry. *J. Appl. Microbiol.* 29: 966-975.

Farber J. M. (1996). An introduction to the hows and whys of molecular typing. *J. Food Prot.* 59:1091–101.

Fuller, R. (1989). Probiotics in man and animals. *J. Appl. Bacteriol.* 66: 365-378.

Gilliland, S.E. and Walker, M. L. (1990). Deconjugation of bile acids by intestinal Lactobacilli. *Appl. Envrn. Microbiol.* 33: 15-18.

Havenaar, R. and Huis, J. H. J. (1992). Probiotics, a general view in the Lactic Acid Bacteria In: LAB in health and disease. Vol.1. ed. B.J.B. Wood, Chapman and Hall, New York. pp 209-248.

Hejkal, T. W. and Gerba, C. P. (1981). Uptake and survival of enteric viruses in the blue crab, *Collinectes sapidus. Appl. Environ. Microbiol.* 41: 207-211.

Lee, Y. K., Nomoto, K., Salminen, S. and Gorbach, S. L. (1999). Handbook of Probiotics. Jhon Wiley and Sons. Inc. New York.

Lee, Yuan Kun (2003). Food production involving Microorganisms and their products. In: Microbial Biotechnology (Principles and Application), 2nd edn. World Scientific Publishing Co. Pvt. Ltd., Singapore. pp 203-206, 318-320.

Likotrafiti, E., Manderson, K. S., Fava, F., Tuohy, K. M. Gibson, G. R. and Rastall, R. A. (2004). Molecular identification and anti-pathogenic activities of putative probiotic bacteria isolated from faeces of healthy elderly individuals. *Microb. Ecol. Health Dis.* 16: 105–112.

Lindgun, S. E. and Dobrogosz, W. Z. (1990). Metabolic activity of Lactic Acid Bacteria. *Microbial rev.* 87: 149-164.

Lofstrom, C., Knutsson, R., Axelsson, C. E. and Radstrom, P. (2004). Rapid and specific detection of *Salmonella* sp. in animal feed samples by PCR after culture enrichment. *Appl. Environ. Microbiol.* 70: 69–75.

Lu, J. Idris, U., Harmon, B., Hofaere, C., Maurer, J. J. and Lee, M. D. (2003). Diversity and succession of the intestinal bacterial community of the maturing broiler chicken. *Appl. Environ.Microbiol.* 69: 6816–6824.

Metchnikoff, Elie (1908). The Prolongation of Life. Putmans Sons, New York. 1: 1-27.

Parker, R.B. (1974). Probiotic, the other half of the antibiotic story. *Anim. Nutrn. Health.* 29: 4-8.

Parvez, S., Malik, K. A., Ah Kang, S. and Kim, H. Y. (2006). Probiotics and their fermented food products are beneficial for health. *J. Appl. Microbiol.* 100(6): 1171-1179.

Prajapati, Jashbhai B. and Nair, Baboo N. (2003). The history of fermented foods. In: Handbook of Fermented Functional Foods, ed. Edward R. Farnwarth, CRC Press, USA. p 31.

Reilly, P. J. A., Parry, R. W. H. and Barile, L. E. (1990). Post-harvest Technology, Preservation and Quality of Fish in South East Asia. International Foundation for Science, Stockholm. pp. 176-177.

Reque, E. F., Pandey, A., Franco, S. G. and Soccol, C. R. (2000). Isolation, identification and physiological study of *Lactobacillus fermentum* for use as probiotic in chicken. *Braz. J. Microbiol.* 31: 57-65.

Sander, M. E. and Huis in't Veld, J. (1999). Bringing a probiotic containing functional food to the market: microbiological, product, regulatory and labeling issues. *Anton. Leeuw. Int. J. G.* 76: 293-315.

Saxelin, M. (1996). Colonization of the human gastrointestinal tract by probiotic bacteria. *Nutr. Today* 31: 5-8.

Shah, N.P. (2001). Functional foods from probiotics and prebiotics. *Food Technol.* 55: 1901-1906.

Smith, G. C., Aurelian, L., Santosham, M. and Sack, R. B. (1983). Rotavirus associated traveler's diarrhea: Neutralizing antibody in asymptomatic infections. *Infect. Immun.* 41: 829-833.

Stanton, C., Gardiner, G., Meehan, H., Collins, K., Fitzgerald, G., Lynch, P. B. and Ross, R. P. (2001). Market potential for Probiotics. *Amer. J. Clin. Nutr.* 73: 476-483.

Steinkraus, K. H. (1983). Lactic acid fermentation in the production of foods from vegetables, cereals and legumes. *Antonie van Leeuwenhoek.* 49: 337-348.

Suverna, V. C. and Boby, V. U. (2005). Probiotics in human health: A current assessment. 88: 1774-1748.

Thapa, N., Pal, J. & Tamang J. P., (2004). Microbial diversity in ngari, hentak, and tungtap, fermented fish products of North-East India. *World J. Microbiol. Biotechnol.*, 20(6), 599-607.

Zhu, X.Y., Zhong, T., Pandya, Y. and Joerger, R. D. (2002). 16S rRNA – based analysis of microbiota from the cecum of broiler chickens. *Appl. Environ. Microbiol.* 68: 124–137.

# Chapter 6

# Starter Culture in Fermenting Fish

*"It is unbelievable that such a tiny streamlet, as was our initial test trial with the first bacterial starters exactly 40 years ago, grew to a big river for an important industry. Who would have believed then that the use of starter cultures would be a matter of course today"* (Niinivaara, 1994). The idea of inoculating *Lactobacillus* into dry sausage material was introduced by Jensen and Paddock in 1940 (US Patent 2,225,783) with the aim to reduce the ripening time as well as ensuring the quality and aroma of dry sausages. The first LAB meat starter culture, introduced as a pure culture of *Pediococcus cerevisiae*, was developed in 1955 in the USA by Niven *et al.* (1955). At the same time, Niinivaara (1955) applied *Micrococcus* M53 to dry sausage production in Europe. The work of Niinivaara was continued by Nurmi (1966) who combined micrococci with *Lactobacillus plantarum*. Today, several companies provide *Lactobacillus* sp., *Pediococcus acidilactici*, *P. pentosaceus*, *Staphylococcus*

*xylosus* or *S. carnosus* strains in pure cultures or their mixtures for fermentation of meat (Hammes *et al.* 1985). A starter culture may be defined as a preparation or material containing large numbers of variable microorganisms, which may be added to accelerate a fermentation process. Being adapted to the substrate, a typical starter facilitates improved control of a fermentation process and predictability of its products (Holzapfel, 1997).

Most traditional fermented products result from natural fermentations carried out under nonsterile conditions. The environment resulting from the chemical composition of the raw materials, fermentation temperature, absence or presence of oxygen, and additives such as salt and spices causes a gradual selection of microorganisms responsible for the desired product characteristics. The main advantage of natural fermentation processes is that they are fitting to the rural situation, since they were in fact created by it. Also, the consumer safety of several fermented foods around the world is improved by lactic acid fermentation, which creates an environment that is unfavorable to pathogenic enterobacteriaceae and bacillaceae. In addition, the variety of microorganisms present in a fermented food can create rich and full flavours that are hard to imitate when using pure starter cultures under aseptic conditions. However, natural fermentation processes tend to be difficult control if carried out at a larger scale; moreover, the presence of a significant accompanying microflora can accelerate spoilage once the fermentation is completed. Particularly with increased holding periods between product fermentation and consumption when catering for urban markets, uncontrolled fermentations under variable conditions will cause unacceptable wastage by premature spoilage. Techniques to stabilize fermentations operating under nonsterile conditions would therefore be appropriate in the control of natural fermentations, which can be achieved by the use of pure culture starters. A different tool to stabilize fermentations under nonsterile conditions is the use of **multi-strain dehydrated starters**, which can be stored at ambient temperatures, enabling more flexibility. Such homemade starters are widely used in several Asian food fermentations. These starters are more homogenous and their dosage is convenient, but because they are manufactured under nonsterile conditions, some are heavily contaminated with spoilage-causing bacteria and yeasts. This requires quality monitoring of the inoculum and of the fermentation process in which it is used.

Modern starter cultures are selected either as single or multiple strains, specifically for their adaptation to a substrate or raw material. Spontaneous fermentations, i.e. process initiated without the use of a starter inoculum, have been applied in fish preservation for millennia and were elucidated through trial and error, perhaps over thousands of years. The majority of small-scale fermentations in developing countries and even some industrial processes are still conducted as spontaneous processes. Various types of starter cultures and even back-slopping are widely used in fermentation processes, even in industrialized countries. Spontaneous fermentations typically result from the competitive activities of a variety of contaminating microorganisms. Those best adapted to the food substrate and to technical control parameters, eventually dominate the process. The production of metabolites such as organic acids, inhibitory to other contaminating microbes (e.g. Enterobacteriaceae) may provide an additional advantage during fermentation. Bacteria typically dominate the early stages of fermentation processes, owing to their relatively high growth rate, followed by yeasts, in substrates that are rich in fermentable sugars. In numerous traditional fish fermentation processes, material from a previous successful batch is added to facilitate the initiation of a new process. Through this practice of back-slopping, the initial phase of the fermentation process is shortened and the risk of fermentation failure in selection of best-adapted strains, some of, which may possess features that are desirable for use as starter cultures. Initiation of a spontaneous fermentation process takes relatively long time (24-48 hours), with high risk for failure. During this phase which is associated with the lag phase of microbial growth, contaminating microorganisms on raw materials, utensils and from the environment, slowly increase in number and compete for nutrients in order to produce metabolites. This phase can be shortened by inoculation either through back-slopping or with the use of selected starter culture.

Fermentation is generally considered as safe and acceptable preservation technology for improving the hygienic quality and safety of foods. Failure of fermentation process can, however, result in spoilage and/or the survival of pathogens, thereby creating unexpected risks in food products, which would otherwise be considered safe. Inoculation with starter cultures does not provide an absolute guarantee against failure of fermentation processes, nor does it eliminate health hazards associated with pathogens, toxinogens, toxic components or residues. Metabolic activities of desirable fermentation microorganisms must be supported by observing the basic principles of Good Manufacturing Practice (GMP). This implies the maintenance and control of technical parameters that ensure the desired outcome of the fermentation process. Precautions should also be taken against the introduction or transfer of potential health hazards or factors that are potentially detrimental to quality during the fermentation process. The Hazard Analysis Critical Control Point (HACCP) system, presents a scientific and systematic approach for enhancing the safety of foods, from primary production to final consumption, through the identification, evaluation and control of hazards that are of significance for food safety (Amoa-Awua *et al.*, 1998; WHO, 1995).

Considerations for applying starter cultures at the household level should take into account cultural traditions, dietary habits and raw materials, which differ across regions and continents. Lactic fermented fish products produced by small-scale spontaneous solid or semisolid-state fermentations are widely accepted and appreciated by consumers in different parts of the world. In Southeast Asia, on the other hand, traditional fermentations are reliant on moulds as the dominant organisms.

From a technical viewpoint, cost/benefit ratios, logistic factors and the willingness of the small-scale processor to accept new approach is critical in any assessment of the feasibility of introducing the use of starter cultures in small-scale fermentations. In addition, a minimum set of standards or quality parameters for the handling and maintenance of these cultures would need to be developed for use at the artisanal level. The introduction of starter cultures should be considered within the context of realistic prospects for: information transfer minimal technical adjustments to small-scale "low-tech" fish fermentations, application of the HACCP system and education and training. The prospect of applying starter cultures will become attractive to the small-scale processor only if benefits, such as reduction of costs (e.g. energy), reduced fermentation times, reduced risk of spoilage (increased shelf-life), improved process control, improved sensory quality (taste, aroma, visual appearance, texture, consistency), improved safety attributes and reduced preparation procedures for the final product, are achieved.

Modern food biotechnology has moved a long way since ancient times of empirical food fermentations. Preservation and safeguarding of food are, however, still major objectives of fermentation. In addition, other aspects, such as wholesomeness, acceptability and overall quality, have become increasingly important and valued features to consumers even in developing countries where old traditions and cultural particularities in food fermentations are generally well maintained. The trend today is food for special health use, referred to as functional food, with the aim to promote the health and well-being of the consumer. An important class of functional foods encompasses probiotics, prebiotics and synbiotics. Probiotics has been described in chapter 5. Prebiotics are non-digestible food ingredients that improve the health status of the consumer through selective stimulation of growth and metabolic activity of a small number of bacteria in the colon, in particular Bifidobacteria. Synbiotics are food products combining probiotics and prebiotics. The expected functionality of probiotic, prebiotic and synbiotic food products offers very attractive perspectives, both from a scientific as well as from a commercial (industrial) point of view. Starter culture aided fermentation in real sense, is a kind of value addition to the finished product. Such starter cultures can be referred to as Functional starter cultures, which have inherent functional properties of nutritional, organoleptical or technological importance. Optimal control and monitoring of the *in situ* production of important metabolites from food grade microorganisms such as lactic acid bacteria could lead to a better understanding of food fermentation processes, make them more controllable, and hence expand the use of functional (multiple) starter cultures in the food industry.

In addition to those lactic acid bacteria (LAB) that are part of the normal flora of human gastrointestinal tract, large numbers of LAB are introduced into the gastrointestinal tract as constituents of foods. However, LAB typically found in food do not survive the gastric conditions and passage through human gastrointestinal tract. Therefore, they may not contribute to the health of the consumer. Probiotic strains used in fermented foods have been selected to withstand these conditions and reach the intestine as viable organisms thus facilitating potential health effects. There are various reports of using different strains of microorganisms as starter culture in the fermentation of a variety of food products. Some LAB and yeast strains associated with fermented foods, are capable of degrading antinutritional factors, such as phytic acid and phenolic compounds. Incorporation of these organisms into starter cultures may, therefore, to serve upgrade the nutritional value of foods. Furthermore, selected strains may enhance the general benefits of spontaneous fermentation such as improved protein digestibility and micronutrient bio-availability, and contribute more specifically to biological enrichment through the biosynthesis of vitamins and essential amino acids.

Speck (1972) reported that the bacteria added to the raw sausages mass rapidly multiply in it. The lactic acid arising from the enzymatic degradation of the muscle glycogen and added sugar lowers the pH to about 5 and below. This pH value inhibits the growth of acid sensitive spoilage organisms. In addition to this preservative action, the lowering of the pH also leads to changes in the consistency of the sausage mass. The isoelectric pH of meat protein is at pH 5.3 and at this value, the protein passes into a gel-like state. The solidity so arising is partially responsible for the firmness of the raw sausage.

Etchells *et al.* (1975) described a controlled fermentation of pickles with the use of starter culture. The washed gherkin are disinfected in a chlorine solution and placed in brine. This is acidified with acetic acid and buffered with sodium acetate. The fermentation is initiated by the addition of starter culture of *Lactobacillus plantarum* and *Pediococcus cerevisae*. Excess $CO_2$ is removed by the introduction of nitrogen. The fermentation is completed after 7-12 days at 25-30$^0$C (instead 0f 3-6 weeks). This method results in increased yields and improves quality. Higher temperatures of fermentation are being used to increase starter culture proliferation and shorten incubation time. But the high temperature is also conducive to the growth of pathogens such as *Staphylococcus aureus*. In dry sausage manufacture where smoking is not common, *Staphylococcus aureus* may be more of a problem if higher fermentation temperature is used. Smith and Palumbo (1981) observed that the use of microflora (starter culture) in production of flavour compounds, antioxidants and antimicrobial compounds other than lactic acid may significantly control such pathogens.

Morzel *et al.* (1997) studied the integration of defined starter culture into fish fermentation process. Six Lactobacilli and five Staphylococci strains were tested in salmon model system. Dependence of bacterial growth and pH reduction of sucrose, sodium chloride and sodium nitrite concentration and fermentation temperature were evaluated. A combination of *Lactobacillus sake* and *Staphylococcus carnosus* were applied to whole salmon fillets to which sodium chloride and sucrose had been added.

The influence of sucrose and sodium chloride concentration, temperature and the addition of sodium nitrite on pH, on colour, and microbial stability were evaluated. Optimal conditions were found to be 25g/kg sodium chloride, 15g/kg sucrose, and fermentation at 12 $^{0}$C for 3 days, subsequent storage at 4 $^{0}$C.

There is evidence that lactic acid bacteria involved in the fermentation of vegetables are also active in meat fermentation. Stamer (1974) divided sauerkraut fermentation into three phases of microbial participation based on predominant microflora type, in contrast to multistage progression in meat. The recession of undesirable flora due to acid production and anaerobiosis, followed by dominance of heterofermentative lactic flora and at the last stage, the aciduric homofermentative populations dominate which stops fermentation. Heterofermentative and homofermentative bacteria both produce lactic acid, but only the heterofermenters contribute such end products as $CO_2$, ethanol, acetic acid and acetaldehyde in amounts that contribute significantly to the flavour and aroma of good sauerkraut. Indeed when the heterofermenters dominate fermentation, as happens when the fermentation temperature is too high; the sauerkraut will taste like acidified cabbage (Vaughn, 1982).

Everson *et al.* (1970) reported that properly selected, physiologically active starter culture will ensure the required pH decrease, safety of sausage as well as improve its uniformity in sense of flavour, appearance and texture and shorten production cycles. In the production of dry sausage, fermented by *Lactobacillus rhamnosus* strains GG, LC-705, E-97800 as well as *Pediococcus pantosaceus* E-90390 and *Lactobacillus plantarum* E-98098, Erkila *et al.* (2001) reported their suitability for use as probiotic starter cultures in fermenting dry sausage with respect to the viable lactic cell, acceptable biogenic amine level as well as flavour profile compared to the commercial starter culture.

A more recent study has reported the **gene fusion vector as starter culture**. Gene fusion vectors were constructed for *Staphylococcus carnosus*, a starter culture for fish and meat product fermentation. Vectors were constructed from Staphylococcal plasmid pC 194 and a *Staphylococcus hyicus* gene encoding a lipase (EC - 3.1.1.3) with unusual properties. The lipase was secreted by *Staphylococcus carnosus*, so its original peptide should be useful for foreign protein secretion. An *E. coli* TEM beta-lactamase (EC - 3.2.5.6) gene was inserted at various positions of the lipase gene, resulting in fusion proteins with varying lengths of lipase portions. The lipase signal peptide alone was not sufficient for efficient secretion of the hybrid protein into the culture medium. Another gene fusion vector, plasmid pTIT2T, was constructed using the *Staphylococcus aureus* Protein–A gene under the control of the phage lambda pR promoter. The lipase gene without the pro-peptide was linked to the Protein-A gene in the correct reading frame. The fusion protein was expressed in *E. coli*, and showed lipase activity (Goetz, *et al.*, 1988).

Most of the indigenous fermented foods are prepared by processes of solid substrate fermentation in which the substrate is allowed to ferment either naturally or by adding starter cultures. In East and South-East Asia, filamentous moulds are predominant microorganisms in the fermentation processes, whereas in Africa, Europe and America, fermented products are prepared exclusively using bacteria or bacteria-yeasts mixed cultures; moulds seem to be little or never used. However, in the Himalayas, all three major groups of microorganism (moulds-yeasts-bacteria) are associated with indigenous fermented foods and beverages (Tamang, 1998), showing the transition food culture. As already stated earlier that the fermentation process of Tungtap is a traditional and uncontrolled system and lack of any scientific influence. The study on the microbiology of partially cured and fermented tungtap revealed a variety of bacteria and fungal members all of which are not desired during fermentation. Also some members were found to be pathogenic as evidence from assessing the GOT and GPT level. Moreover, the off flavour produce at the end of fermentation in the traditionally fermented fish is one of the reasons why this product although nutritionally rich, confined only within the geographic boundary.

In recent times there has been considerable focus on the inclusion of mycotoxin-degrading strains in starter cultures. The use of mycotoxin contaminated raw materials for fermentation in developing countries, poses a special challenge for the selection of strains that are capable of mycotoxin detoxification (Adegoke *et al.*, 1994; Smith *et al.*, 1994; Westby *et al.*, 1997; Holzapfel *et al.*, 1998).

Spontaneous fish fermentations are neither predictable nor controllable. Pure cultures isolated from mixed populations of traditional fermented foods exhibit a diversity of metabolic activities, which vary even among strains. These include differences in growth rate, adaptation to a particular substrate, ability to degrade antinutritive factors, antimicrobial properties, flavour and quality attributes and competitive growth behaviour in mixed cultures (Holzapfel, 1997). Single- and mixed-strain cultures must, therefore, be tested at the pilot scale, before their use in small-scale operations.

The introduction of starter cultures in traditional small-scale fish fermentations should incorporate considerations for improving processing conditions and product quality through: (1) rapid accelerated metabolic activities (acidification); (2) improved and more predictable fermentation processes; (3) desirable sensory attributes; (4) improves safety and reduced hygienic and toxicological risks. Commercial starter cultures generally originate either from food substrates or from the processes in which they are applied. Environmental conditions, back-slopping, adaptation and the repeated use of specific fermenting vessels can contribute to the selection of microbial populations typical of a fermentation process. The selection of suitable starter strains should take into account their interactions in mixed cultures, with consideration for the behaviour of these strains under defined conditions, and within the food substrate. Other factors, which should be considered, include – (1) competitive behaviour, viability and survival; (2) antagonism against pathogens and spoilage microbes; (3) the rate of acid production; (4) organoleptic changes; (5) primary metabolites of fermentation; (6) degradation of antinutritive factors; (7) detoxification; (8) probiotic features. Modern approaches incorporate considerations for technical safety and health-promoting features in the selection of the most optimal strain(s) for a process. Ideally a multifunctional strain should be the target.

Technical aspects of starter culture development should incorporate considerations relevant to adoption of the starter to the substrate, the rate of acid production, fermentation metabolites (e.g. hetero-vs. homo-fermentation) and the ability of single or mixed strain cultures to produce desirable sensory qualities in the fermented product. Numerous reports indicate that

*Lactobacillus brevis, L. fermentum, L. plantarum, L. reuteri, Pediococcus pentosaceous* and *P. acidilactici* exhibit superior performance in lactic fermented food products (Steinkraus, 1996; Holzapfel, 1997; Lee, 1997; Oyewole, 1997). Several LAB are associated with meat and fish fermentations. *L. sakei* and *L. curvatus* (Hammes and Hertel, 1998) have been determined to be superior starter cultures for meat fermentations. Acidification to pH values of less than 4.2 constitutes a major safety concern in LAB aided fermented foods. Recent observations, however, confirm that a number of metabolites, such as acetic acid (from heterofermentative LAB), hydrogen peroxide and bacteriocins, produced during fermentation process, exhibit antimicrobial properties which may contribute to the safety of lactic fermented fish and other food products. Moreover, organic acids, which show strong antagonistic effects in the undissociated form at lower pH values, are particularly effective in inhibiting Gram-negative bacteria, such as pathogens.

Bacteriocins are antimicrobial substances of proteinaceous nature that are active against closely related bacteria. They exhibit a narrow of activity but are not active against Gram-negative bacteria. Proteolytic enzymes present in food substrate are capable of inactivating bacteriocins. Bacteriocins from LAB may be classified into three structural groupings on the basis of their physico-chemical and antimicrobial properties (Schillinger et al., 1995; Holzapfel *et al,* 1995). Bacteriocinogenic LAB strains are therefore, of special interest in view of their possible application in food safety assurance. Bacteriocinogenic LAB have been shown to effectively inhibit the growth of pathogens, such as *Listeria monocytogenes, staphylococcus aureus, Bacillus cereus* and *Clostridium dificile,* even under *in situ* conditions (Holzapfel *et al,* 1995).

The **lactic acid isomer** produced during fermentation is typically related to the LAB species from which it is produced. LAB genera such as *Streptococcus, Lactococcus, Enterococcus,* and *Carnobacterium* produced > 90% of the L (+)- isomer as an end product of sugar fermentation. *Leuconostoc* spp. and *L. delbrueckii* (all sub species) on the other hand produced D(-)- lactic acid. Lactic acid isomers by lactobacilli and pediococci are species specific. The L (+)- isomer is produced by *L. casei,* for example, while a recemate (DL) is produced for *L. sakei,* all heterofermentative lactobacilli and practically all *Weissella* spp. The nature of lactic isomer is of concern, since high levels of the D(-)- lactic acid isomer are not hydrolysed by LDH enzymes in humans and are, thus capable of causing acidosis. WHO recommendations indicate a maximum of this daily intake of 100mg/Kg body weight of this non-physiological lactic acid isomer (WHO, 1968). There are, however, no recommended limitations for intake of L(+)- lactic acid isomer.

Biogenic amines are frequently produced by amino acid decarboxylase positive microorganisms, during fermentation. The occurrence of biogenic amines in traditional fermented fish products has, however, been reported (Jae-Hyung *et al.,* 2002). Certain LAB such as *L. buchneri,* have been shown to produce biogenic amines, such as histamine, putrescine, tyramine and cadaverine, in fermented products of plant and animal origin. A number of bacteria associated with these fermentations, however, exhibit the potential for degrading histamine and tyramine through the production of mono- and di-amino-oxidases (Leuschner *et al.,* 1998).

Experienced gained in the field of traditional fermentation technologies has shown that the process may be accelerated through the addition of a starter obtained from previous fermentation batch (back-slopping) as in traditional back-slopping, inoculum from a previous batch of fermented product contains large numbers of desirable microorganisms in an active state, which are adapted to the substrate. Inocula consisting of a portion of a fermenting substrate may be preserved by dehydration (air-or sun-drying) and grinding into a powder. Dehydration enhances the viability of microorganisms over relatively long periods, provided the product is maintained in the dehydrated state. Natural preservation of microorganisms can also be accomplished with the use of a carrier, such as the fermentation vessels or an "inoculation belt", such as that used in Ghana for the initiation of '*Pito*' beer fermentations. Although such preservation methodologies are common to many regions and probably have along tradition, they have not been adequately studied. But these old traditions in starter preparation, preservation and distribution present an extremely valuables basis for the development and application of the other types of starters in small-scale processing, and a number of attractive challenges and benefits to the entrepreneur: (i) the use of simple, inexpensive vessels that are readily available; (ii) flexibility and simplicity of maintenance and handling; (iii) minimal loses due to fermentation failure; (iv) job creation and income generation in rural areas; (v) handling and storage at household level; (vi) distribution and sale in local markets.

Spontaneous fermentations typically result from the competitive activities of different microorganisms. Strains best adapted and stages of the process. The complexity and variability of microbial populations associated with these fermentations is somewhat reduced in back-slopping operations, where processing conditions and continued recycling of a portion of a previous batch, determines dominance of the best adapted strains. Inoculation with a single-strain culture can eventually result in a mixed-strain fermentation if raw materials are not sterilized prior to inoculation and maintained axenic (free from foreign microorganisms) throughout strict process control.

Single-strain cultures offer advantages of improving both process control and the predictability of metabolic activities within the cultures. They are, however, relatively easily degraded by bacteriophage infection, spontaneous mutation or through the loss of key physiological properties (e.g. plasmid-mediated fermentation of lactose). Deterioration in culture performance due to one or more of these effects adversely affects the fermentation process. Modern equipment for the preparation, handling and application of pure single-strain cultures and for strict process control at all stages of the fermentation, are not available or attainable in most small-scale operations. In large-scale fermentations, on the other hand, consistent end product quality is achieved over extended time periods through the use of defined single-strain cultures and properly controlled processes.

Mixed strain cultures, such as those of the *tungtap* type, on the other hand, are less susceptible to deterioration and are, thus, better suited to most small scale operations. Mixed strain cultures are relatively unaffected by fluctuation conditions of handling, storage and applications. In addition, they contribute to a more complex sensory quality, whilst producing favourable synergistic effects, such as the degradation of undesirable factors, flavour production and accelerated ripening and maturation. Variation in product quality with the use of mixed strain cultures can be minimized through proper process control.

Plant materials containing fermentable sugars provide suitable substrate for yeast species of *Saccharomyces*, *Candida*,

*Torula*, *Hansenula* and others. Such yeasts species are associated with traditional fermented fish products of India. LAB are major importance among bacteria associated with traditional fermented foods as mentioned in the previous chapter. The largest spectrum and richest variety of lactic fermented foods is probably found in Africa. The association of LAB with the human environment and their beneficial interactions, both in food and in the human intestinal tract, combined with the historic tradition of lactic fermented foods in many cultures, has led to the conclusion that these foods may be recognized as safe. Several factors must be taken into consideration when evaluating the use of LAB as starters - (a) not all LAB are of equal technical and practical importance in fish or other food fermentations, (ii) *Lactobacillus* (both homo- and heterofermentative), *Leuconostoc* and, to a lesser extent, *Pediococcus*, *Lactococcus*, *Enterococcus* and *Weissela* are the genera which generally occur in traditional fermented foods, (iii) the genus *Bifidobacterium*, although phylogenetically not related to LAB, is often grouped as part of the LAB for its probiotic functions (Holzapfel *et al.*, 1998), (iv) with the exception of *S. thermophillus*, species of the genus *Streptococcus* are generally regarded as pathogens, (v) Suitable cultures for fermentation must be selected at strain level since no all strains of a species are equally suitable for use as starters, nor are all equally well adapted to a food substrate.

The association of certain strains of *Enterococcus faecium*, *E. faecalis* and *L. rhamnosus*, with exceptional cases of endocarditis, should be no absolute reason to disqualify the use of food grade strains of these and other LAB species from their potential use in food fermentations and even as probiotics. LAB starters are not yet commercially available to the small-scale fermentation of traditional African foods and fish product in India. Making them available to the small-scale processor, however, poses a great challenge to both the food microbiologist and the potential entrepreneur. These starter cultures may find application and might serve to improve small-scale fermentations even in rural areas, upon identification and selection of suitable strains. Such approaches will, however, only be successful if the basic principles of good processing (GMP) are observed.

In the traditional system of fermentation of the fish, *Puntius sarana,* the semi-cured or the partially cured fish is used. Therefore, the bacterial members which are present in the partially cured fish are coming from their habitat and also associated during curing process. Similarly, the bacteria present in the fermented fish i.e. tungtap are the members from partially cured fish which could thrive the acidic condition during fermentation, and also associated from the earthen pot used for fermentation process as the traditional system of fermentation follow 'back sloping'. Therefore, in selecting the starter culture for fermenting this fish, the bacterial isolates from the partially cured fish can be the candidate of choice.

Certain criteria were followed while selecting the good candidature in formulating the starter culture. These include screening for bacterial enzymatic profile with low lipolytic and proteolytic activity and moderately high amylolytic activity. The High lipolytic and proteolytic activity causes high rancidity and oxidation of fats contributing to the characteristic off flavour in the finished product and therefore, unwanted. On the other hand bacteria with a moderately high amylolytic activity are essential as it helps in liquefaction during processing of fish product. Secondly, antibacterial activity against certain pathogenic organisms was another criterion for selecting the possible starter culture in fermenting fish. Both partially cured and the traditionally fermented tungtap contains certain bacterial isolates which are found to be pathogenic in nature although present in small numbers. Thirdly, probiotic characterization of LAB is probably the most important criterion as the probiotic bacteria can exert a health benefit to those consumers who prefer the fish product to take as it is after fermentation without cooking or frying. Moreover, the antimicrobial activity of the probiotic bacteria can eliminate the chances of contaminating with any pathogenic or toxigenic microbes if present in the sample. Fourthly, the antagonism among microbes was another criterion for selecting starter culture. Because, some microorganisms shows antagonism against a particular organism while some other shows mutual growth. From the study two Lactobacillus species viz. *Lactobacillus fermentum* (BI26) and *Lactobacillus lactis* (BVS14) isolated from intestine and skin of partially cured fish respectively were found to be suitable to be used as starter culture in fermenting fish *Puntius sarana.*

*Preparation of Raw-materials for Fermentation :* The raw-material i.e. the partially cured fish along with the oil (refined oil was used instead of fish oil as in traditional system) and the clay pots were packed in polythene bags and autoclaved at 15lb pressure for 15 minutes. Lactose as fermentable sugar, which was used in the fermentation process was also sterilized in autoclave at 12lb pressure for 10 minutes. The reason behind sterilization was to eliminate the microbes present in all the ingredients and to carry out fermentation solely by the starter culture added either singly or in combination.

*Microbial Analyses of Sterilized Partially Cured Fish, Oil and Clay Pots :* The sterilized partially cured fish and some pieces of clay pots were inoculated into nutrient broth and incubated at 37 $^{0}$C for 24 hours. Later, from the respective broth one loopful of 24 hour old culture was streaked onto the nutrient agar plates and incubated at 37 $^{0}$C for 24 hours, while the refined oil was directly streaked onto the nutrient agar plates, in order to verify the microbial load, if persists, in these materials. The sterilized partially cured fish and other ingredients after autoclaving were examined for presence of any microbes and they were found to be free of any microbial load.

*Physicochemical and Biochemical Analyses of Sterilized Partially Cured Fish :* The changes in physicochemical and biochemical profile of the partially cured fish after autoclaving, if any, were analyzed. Analysis on moisture content, pH, crude protein, amino acid content, carbohydrate content and reducing sugar content were mainly carried out. The physicochemical and biochemical studies of the partially cured fish sample after sterilization have shown no variation only except with a little increment of protein content from 44.56% in unsterilized sample to 44.58%; and amino acid content from 0.05mg/ml in unsterilized sample to 0.08mg/ml.

*Microbiological Analyses of Starter Culture added Fermented Fish Product:* After 10 days, 20 days and 30 days of fermentation, one clay pot from each set of experiments were studied for microbial load followed by colony and microscopic study and finally various biochemical tests were performed to identify them using Bergey's Manual of Determinative Bacteriology (Buchanan, 1975). The examination for bacterial load in the products fermented with *Lactobacillus fermentum* and *Lactobacillus lactis* in singly or in combination showed lowest bacterial load of 0.6 x $10^6$ c.f.u/ml (at $10^{-4}$ dilution) after 10 days of fermentation in sample fermented with only *L. lactis* while it was recorded to be highest in case of sample fermented with both the bacteria and was recorded to be 1.07 x $10^6$ c.f.u/ml (at $10^{-4}$ dilution). After 20 days of fermentation the bacterial load was found to increase and after 30$^{th}$ day of fermentation the highest bacterial load was recorded in case of sample fermented with both the bacteria (1.97 x $10^6$ c.f.u/ml at $10^-$

$^{4}$ dilution) followed by sample fermented with *L. fermentum* with a bacterial load of 1.88 x $10^7$ c.f.u/ml (at $10^{-5}$ dilution) and the lowest bacterial load was recorded in case of sample fermented with *L. lactis* with a bacterial load of 0.67 x $10^7$ c.f.u/ml at $10^{-5}$ dilution (Table 17). The lactic acid bacteria usually grow in a sequence in a food product. Generally, the Leuconostocs or Streptococci begin the fermentation and are followed by Pediococci and Lactobacilli (Banwart, 1987). This may be reason why the bacterial load was initially low and continue to increase as the day of fermentation proceeds.

**Table 17.** Changes in bacterial load of fish samples fermented with starter culture.

| Day of Ferment-ation | Sample Fermented with *Lactobacillus fermentum* | Sample Fermented with *Lactobacillus lactis* | Sample fermented with *Lactobacillus fermentum* + *Lactobacillus lactis* |
|---|---|---|---|
| 10 | 0.8 x $10^6$ (at $10^{-4}$ dilution) | 0.6 x $10^6$ (at $10^{-4}$ dilution) | 1.07 x $10^6$ (at $10^{-4}$ dilution) |
| 20 | 1.21 x $10^6$ (at $10^{-4}$ dilution) | 1.1 x $10^6$ (at $10^{-4}$ dilution) | 1.53 x $10^6$ (at $10^{-4}$ dilution) |
| 30 | 1.88 x $10^7$ (at $10^{-5}$ dilution) | 0.67 x $10^7$ ) (at $10^{-5}$ dilution | 1.97 x $10^7$ (at $10^{-5}$ dilution) |

The fish products were harvested after 10$^{th}$, 20$^{th}$ and 30$^{th}$ day of fermentation. Bacteria were isolated and identified on the basis of the colony morphology, microscopic characters and various biochemical tests. The findings revealed that the bacterial isolates used as starter cultures for fermentation in each set of experiment could only be recovered and no other bacterial isolates were found in any of the preparation indicating that the samples were not contaminated by undesired organisms of the surrounding environment and that the fermentation was carried out solely by the starter cultures itself.

The fish products fermented with the two starter cultures were also examined for their sensory evaluation which showed that initially the texture was stiff and it became soft in the final product i.e. at 30th day of fermentation. This may be due to the presence of various metabolites from bacteria that soften the bones of the fish thereby enhancing palatability during the process of fermentation. The colour of the fish has been changed from silvery to light brown to brownish and finally dark brown on 30th day of fermentation (Table 18).

**Table 18.** Sensory evaluation of the fish during the time of fermentation.

| Days of Fermentation | Texture | Aroma | Colour |
|---|---|---|---|
| 0 | Stiff | Fishy | Silvery |
| 10 | Soft with little Stiffness | Fishy | Light Brown |
| 20 | Soft | Fishy | Brownish |
| 30 | Soft | Fishy | Dark Brown |

*Physicochemical and Biochemical Analyses of the Fermented Fish Products with Starter Culture:* From all the experimental sets different physicochemical and biochemical parameters like moisture content, pH, titrable acidity, ash content, crude protein, total carbohydrate, reducing sugar, amino acid content, etc. were analyzed at intervals of every 10 days from the days fermentation upto 30 days. The physicochemical analyses of the fish product after 30th day of the fermentation carried out with starter culture in different set of experiments showed the moisture content of 15.02% which is lower than what was found in case of traditionally fermented fish. Similarly, the titrable acidity and pH were recorded to be 1.51% and 5.2 respectively in all the starter culture added samples. The ash content has been slightly increased to 30.43% (Table 19) which, was recorded to be 30.32% in case of traditionally fermented fish sample. The biochemical analyses of these starter culture aided fish products revealed a very high protein content of 56.34% and low lipid content of 2.11%. The product also showed an increase in the total free amino acids than that of traditionally fermented fish. The low moisture content ensures the preservation of the product. Agrhar-Murugkar and Subbulakshmi (2006) reported that the fat content of the fermented food products is higher than the unfermented counterparts. But in the present findings a low fat content was recorded while the protein content in the starter culture added products was found to be higher. The protein content in the products fermented with both the starter cultures in combination was found to be slightly higher than when the starter cultures were used individually (Table 20). The reason can be attributed to the microbes present in the fish which play an important role during the course of fermentation. And when both the bacterial isolates were used in combination the contribution of the microbial protein may lead to the increment of protein in this product. Therefore, the physicochemical and biochemical profiles in the present study suggests that the nutritional quality of the product has not deteriorated while trying to enrobe the process of fermentation in a scientific way rather it increases the quality of the product in terms of its nutritious property and health benefit.

**Table 19.** Physicochemical analyses of fish (*P. sarana*) fermented with starter cultures.

| Days of Fermen-tation | Moisture Content (%) | | | Ash content (%) | | | Titrable Acidity (%) | | | pH | | |
|---|---|---|---|---|---|---|---|---|---|---|---|---|
| | I | II | III | I | II | III | I | II | III | I | II | III |
| 0 | 12.33 | 12.33 | 12.33 | 29.00 | 29.00 | 29.00 | 3.32 | 3.32 | 3.32 | 5.9 | 5.9 | 5.9 |
| 10 | 13.18 | 13.12 | 13.14 | 28.70 | 29.00 | 29.00 | 3.84 | 3.84 | 3.84 | 5.8 | 5.8 | 5.8 |
| 20 | 14.56 | 14.44 | 14.52 | 29.62 | 29.20 | 29.59 | 4.33 | 4.33 | 4.33 | 5.4 | 5.4 | 5.4 |
| 30 | 15.02 | 14.87 | 14.99 | 30.43 | 30.36 | 30.43 | 4.51 | 4.51 | 4.51 | 5.2 | 5.2 | 5.2 |

I = *Lactobacillus fermentum*; II = *Lactobacillus lactis*; III = *L. fermentum* + *L. lactis*

**Table 20.** Biochemical analyses of fish (*P. sarana*) fermented with starter cultures.

| Days of Ferment -ation | Protein Content (%) | | | Total Carb ohyd rate (%) | Reducing Sugar (%) | | | Total Lipid (%) | | | Total Free Amino Acid (mg/ml) | | |
|---|---|---|---|---|---|---|---|---|---|---|---|---|---|
| | I | II | III | | I | II | III | I | II | III | I | II | III |
| 0 | 44.53 | 44.54 | 44.54 | - | 0.04 | 0.04 | 0.04 | 11.02 | 11.02 | 11.02 | 0.08 | 0.08 | 0.08 |
| 10 | 45.00 | 45.00 | 45.05 | - | 0.03 | 0.03 | 0.03 | 6.88 | 6.88 | 6.79 | 0.09 | 0.09 | 0.09 |
| 20 | 48.00 | 48.00 | 48.20 | - | 0.02 | 0.02 | 0.02 | 4.53 | 4.56 | 4.53 | 0.11 | 0.11 | 0.11 |
| 30 | 55.96 | 55.87 | 56.34 | - | 0.02 | 0.02 | 0.02 | 2.11 | 2.13 | 2.11 | 0.13 | 0.12 | 0.13 |

I = *Lactobacillus fermentum*; II = *Lactobacillus lactis*; III = *L. fermentum* + *L. lactis*

*Determination of Water Activity ($A_w$) of the Starter Culture Aided Fish Products, Partially Cured Fish and Traditionally Fermented Fish:* The partially cured fish, traditionally fermented fish and the fish fermented by *Lactobacillus fermentum* and *Lactobacillus lactis* either singly or in combinations were tested for determination of *water* activity. The water activities ($A_w$) of the finished products of the laboratory treated preparations ranged from 0.812 to 0.913 (Table 21), indicating an environment supportive of microbial growth. The water activity of partially cured fish comes within the limit of that of laboratory preparation, whereas it is 0.798 in case of traditionally fermented fish. All these values of water activities may support microbial growth. Since the traditional fermented fish is a product of uncontrolled fermentation, the risk of presence of undesirable organisms can not be ruled out. The water activities of the laboratory prepared products are supportive of microbial growth, but since they were fermented with starter cultures, the finished products contain predominantly of the starter culture. Therefore, undesirable organism, if introduced, maybe eliminated by the members of the starter cultures through competitive inhibition.

**Table 21.** Determination of water activity (Aw) of the partially cured, traditionally fermented and starter culture aided fish products.

| Fish Product | Water Activity ($A_w$) |
|---|---|
| Partially Cured | 0.858 |
| Traditionally Fermented | 0.798 |
| Fermented with *Lactobacillus fermentum* | 0.812 |
| Fermented with *Lactobacillus lactis* | 0.913 |
| Fermented with *Lactobacillus fermentum*+ *Lactobacillus lactis* | 0.834 |

From this investigation it is seen that a little scientific input in the process of fermentation can bring about a desirable change in the finished product with a quality which is much improved if compared with the traditional one. Moreover, a complete controlled environment during fermentation and use of sterilized materials is the added advantage to label it as safe for consumption. The use of two probiotic bacterial isolates for fermentation is expected to be sufficient to influence the consumers as they are always rushing towards functional food. The reduction of off flavour in the starter culture aided product to a great extent can be the reason of their acceptance by the people other than producer community as evidence from the organolaptic test. Therefore, with such a fish product fermented with standard starter culture, the market potential is expected to increase many fold.

No matter how much research is carried out on improved traditional processes or novel products, the ultimate aim is implementation. Unfortunately, a wide gap exists between research data published in scientific journals and the practice of food processing. Much attention should be given to the extent of usefulness of new products to the end user. To this effect, not only should the sensory, nutritional, and other quality characteristics of newly developed products or processes be taken into account, but they should also be integrated with sound price calculations, market surveys, and extension efforts. Only a competitive process has good chances of being implemented.

In this direction, the efforts of TIFAC (Technology Information, Forecasting & Assessment Council) have shown fruitful from the encouraging results obtaining from various projects. Set up in 1988 for keeping technology watch on priority areas and to promote actions, TIFAC is an autonomous society under the Department of Science & Technology (DST), Govt. of India. To ensure rapid socio-economic development, application of science and technology should be the foremost priority. In this background, TIFAC has initiated various projects on the development of value added aquatic products in different geographical areas of India with different women cooperatives in collaboration with fisheries institutes as well as state fisheries departments; who are providing the technical and logistical support to the projects. The objectives of the projects are to establish a commercial level cottage industry by organizing women fishers and to improve upon their socio-economic status; to improve the utilization of low value fish/ shellfish by value added product development; improvement in quality & quality testing of the product; and market linkage.

At present, the new emphasis on food safety will ensure the extensive supplies of fermented food demanded by consumers in various part of the world. Indigenous fermented foods or beverage offer the opportunity for industrial scale-up to market to a much wider population. Recent developments in biotechnology can facilitate this industrialization and lead to a more standardized product with unique characteristics and improved quality along with better storage properties and enhanced nutrition value. Even though changes will be very slow particularly with some products of low levels of exports. For the products of high potential to be exported, e.g., fish sauce, fermentation processes have been developed very successfully to commercial scale through experiences over many years.

## References

Adegoke, G.O., Otumu, E.J., Akanni, A.O. (1994). Influence of grain quality, heat and processing time on the reduction of aflatoxin $B_1$ levels in tuwo and ogi: two cereal-based products. *Plant Foods Hum. Nutr.* 45: 113–117.

Amoa-Awua, W.K., Halm, M., Jakobsen, M. (1998). HACCP System For Traditional African Fermented Foods: Kenkey. WAITRO, 1998, Taastrup, Denmark. ISBN: 87-90737-02-4.

Banwart, G. J. (1987). Useful Microorganisms, In: Basic Food Microbiology. CBS Publishers & Distributors, New Delhi. pp 453, 482.

Buchanan, R. E. and Gibbons, N. E. (1975). Bergey's Manual of Determinative Bacteriology, 8th edn. The William and Wilkins Company, Baltimore.

Erkila, S., Snihko, M. L., Eerola, S., Petaja, E. and Mattila-Sandholm, T. (2001). Dry fermented sausages by *Lactobacillus rhamnosus* strains. *Int. J. Food Microbiol.* 64: 205-210.

Etchells, J. L, Bell, T. A., Fleming, H. P., Kelling, R. E. and Thompson, R. L. (1975). Suggested procedure for the controlled fermentation of commercially brined pickling cucumbers- The use of starter cultures and reduction of carbon dioxide accumulation. *Pickle Pak. Sci.* 3: 4-14.

Everson, C. W., Danner, W. E., Hammes, P. A. (1970). Bacterial starters in sausage products. *J. Agr. Food Chem.* 18(4): 570-571.

Goetz, F., Liebl, W. and Knorr, W. (1988). Gene fusion vectore in *Staphylococcus aureus* as starter culture for fish, DECHEMA-Biotecnol. Conf. pp 267-273.

Hammes, W., Rölz, I. and Banteon, A. (1985). Microbiologische Untersuchung der auf demdeutschen Markt vorhandenen Starterkulturpräparate für die Rohwurstbereitung. Fleischwirtsch. 65: 629-636.

Hammes, W.P., Hertel, C. (1998). New developments in meat starter cultures. *Meat Sci.* 49: (Suppl. 1), S125- S138.

Holzapfel, W.H., Geisen, R., Schillinger, U. (1995). Biological preservation of foods with reference to protective cultures, bacteriocins and food-grade enzymes. *Int. J. Food Microbiol.* 24: 343- 362.

Holzapfel, W.H. (1997). Use of starter cultures in fermentation on a household scale. *Food Control* 8: 241–258.

Holzapfel, W. H., Haberer, P., Snel, J., Schillinger, U., Huis in't Veld, J.H.J. (1998). Overview of gut flora and probiotics. *Int. J. Food Microbiol.* 41: 85- 101.

Jensen, L. and Paddock, L. (1940). US Patent 2,225,783. In: Zeuthen, P. 1995. Historical aspects of meat fermentations. In: Fermented meats. Campbell-Platt, G. and Cook, P.E. (eds.). Blackie Academic and Professional, England. pp. 53-68.

Lee, C.H. (1997). Lactic acid fermented foods and their benefits in Asia. *Food Control.* 8: 259- 269.

Leuschner, R.G., Heidel, M., Hammes, W.P. (1998). Histamine and tyramine degradation by food fermenting microorganisms. *Int. J. Food Microbiol.* 39: 1- 10.

Morzel, M., Fransen, N. G. and Arenat, E. K. (1997). Defined starter culture used for fermentation of salmon fillets. *J. Food Sci.* Chicago, 62: 1214-1217.

Murugkar, D. Agrahar and Subbulakshmi, G. (2006). Preparation techniques and nutritive value of fermented foods from the Khasi tribes of Meghalaya. *Ecol. Food Nutr.* 45: 27-38.

Niinivaara, F. (1955). Über den Einfluss von Bacterienreinkulturen auf die Reifung und Umrötung der Rohwurst. Avta Agr. Fenn. 84. Helsinki. p128

Niinivaara, F. (1994). Geschichtliche Entwicklung des Einsatzes von Starterkulturen in derFleischwirtschaft. In: 1. Stuttgarter Rohwurstforum. Ed. H. Buckenhüskes. Gewürzmüller. p.9-20.

Nurmi, E. (1966). Effect of bacterial inoculations on characteristics and microbial flora of dry sausage. Acta Agr. Fenn. 108. Helsinki. p 77.

Oyewole, O.B. (1997). Lactic fermented foods in Africa and their benefits. *Food Control.* 8: 289–297.

Schillinger, E., Geisen, R., Holzapfel, W.H. (1995). Bakteriozine von milchsa¨urebakterien-eigenschaften, wirkungsmechanismen und genetik. *Biospektrum.* 1: 59- 66.

Smith, J. L. and Palumbo, S. A. (1981). Microorganism as food additives. *J. Food Protect.* 44: 936-955.

Smith, J.E., Lewis, C.W., Anderson, J.G., Solomons, G.L. (1994). Mycotoxins in human nutrition and health. European Union Directorate-General XII, Report 16048EN.

Speck, M. L. (1972). Control of food borne pathogens by starter cultures. *J. Dairy Sci.* 55: 1019-1023.

Stamer, J.R. (1974). Fermented vegetables. In: Fermented foods, Current science and technology. New York, State Agricultural Experimental Station. Special report No. 16.

Steinkraus, K.H. (1996). Handbook of Indigenous Fermented Foods, 2nd edn. Marcel Dekker, New York Revised and Expanded.

Tamang, J.P. (1998). Role of microorganisms in traditional fermented foods. *Indian Food Industry* 17 (3): 162-167.

Vaughn, R.H. (1982). Lactic acid fermentation of cabbage, cucumber olives and other products. In: Prescott and Dunn's Industrial Microbiology. ed. G. Reed, AVI. Westport, Conn., pp 185-236.

Westby, A., Reilly, A., Bainbridge, Z. (1997). Review of the effect of fermentation on naturally occurring toxins. *Food Control.* 8: 329- 339.

WHO (1968). Food Additives 29: 144–148.

WHO (1995). Hazard analysis critical control point system: concept and application. Report of a WHO Consultation with the Participation of FAO, 29- 31 May 1995. WHO/FNU/FOS/ 95.7.

# Author Index

# Subject Index

## B



## C



## D



## E



## F

FSC
www.fsc.org
MIX
Papier | Fördert
gute Waldnutzung
FSC® C083411

Zeitfracht Medien GmbH
Ferdinand-Jühlke-Straße 7
99095 Erfurt, Deutschland
produktsicherheit@kolibri360.de